SpringerBriefs in Molecular Science

Chemistry of Foods

Series Editor

Salvatore Parisi, Al-Balqa Applied University, Al-Salt, Jordan

The series Springer Briefs in Molecular Science: Chemistry of Foods presents compact topical volumes in the area of food chemistry. The series has a clear focus on the chemistry and chemical aspects of foods, topics such as the physics or biology of foods are not part of its scope. The Briefs volumes in the series aim at presenting chemical background information or an introduction and clear-cut overview on the chemistry related to specific topics in this area. Typical topics thus include:

– Compound classes in foods—their chemistry and properties with respect to the foods (e.g. sugars, proteins, fats, minerals, …)
– Contaminants and additives in foods—their chemistry and chemical transformations
– Chemical analysis and monitoring of foods
– Chemical transformations in foods, evolution and alterations of chemicals in foods, interactions between food and its packaging materials, chemical aspects of the food production processes
– Chemistry and the food industry—from safety protocols to modern food production

The treated subjects will particularly appeal to professionals and researchers concerned with food chemistry. Many volume topics address professionals and current problems in the food industry, but will also be interesting for readers generally concerned with the chemistry of foods. With the unique format and character of SpringerBriefs (50 to 125 pages), the volumes are compact and easily digestible. Briefs allow authors to present their ideas and readers to absorb them with minimal time investment. Briefs will be published as part of Springer's eBook collection, with millions of users worldwide. In addition, Briefs will be available for individual print and electronic purchase. Briefs are characterized by fast, global electronic dissemination, standard publishing contracts, easy-to-use manuscript preparation and formatting guidelines, and expedited production schedules.

Both solicited and unsolicited manuscripts focusing on food chemistry are considered for publication in this series. Submitted manuscripts will be reviewed and decided by the series editor, Dr. Salvatore Parisi.

To submit a proposal or request further information, please contact Tanja Weyandt, Publishing Editor, via tanja.weyandt@springer.com or Dr. Salvatore Parisi, Book Series Editor, via drparisi@inwind.it or drsalparisi5@gmail.com

More information about this subseries at http://www.springer.com/series/11853

Alessandra Pellerito · Ralf Dounz-Weigt ·
Maria Micali

Food Sharing

Chemical Evaluation of Durable Foods

 Springer

Alessandra Pellerito
Food Safety Consultant
Palermo, Italy

Ralf Dounz-Weigt
Food Sharing Expert
Magdeburg, Germany

Maria Micali
Industrial Consultant
Santa Margherita Marina, Messina, Italy

ISSN 2191-5407 ISSN 2191-5415 (electronic)
SpringerBriefs in Molecular Science
ISSN 2199-689X ISSN 2199-7209 (electronic)
Chemistry of Foods
ISBN 978-3-030-27663-8 ISBN 978-3-030-27664-5 (eBook)
https://doi.org/10.1007/978-3-030-27664-5

This Springer imprint is published by the registered company Springer Nature Switzerland AG
The registered company address is: Gewerbestrasse 11, 6330 Cham, Switzerland

Contents

About the Authors

Alessandra Pellerito is a Biologist graduated at the University of Bologna, Italy (2013), with full marks (110/100 cum laude) after the initial B.Sc. in Biology (Palermo, Italy). After a short period spent in the UK, Dr. Pellerito moved to Germany (Magdeburg), where she has been working for two years as Research Assistant at the Leibniz Institute for Neurobiology. At present, Dr. Pellerito works as a food consultant in the private sector (Germany). Alessandra's first article on food chemistry—Mania I., Barone C., Pellerito A., Laganà P., Parisi S., *Trasparenza e Valorizzazione delle Produzioni Alimentari. L'etichettatura e la Tracciabilità di Filiera come Strumenti di Tutela delle Produzioni Alimentari*, Industrie Alimentari (2017)—concerns authenticity problems, traceability, and food labelling in the current European market.

Ralf Dounz-Weigt is a social worker in case management at the Department of Children's and Young People's Affairs of the city of Magdeburg, Germany. After his B.A. in social works, he is currently studying international social works (M. Sc.) in Hildesheim, Germany. Ralf has gained his experience in food sharing activities working as head of the social facility and project manager at EMMA Spielwagen e.V., Magdeburg in the last six years.

Maria Micali is an experienced author in the field of food science and technology, with a particular focus on chemistry, microbiology, and hygiene. Dr. Micali obtained a Ph.D. in food hygiene from the University of Messina, Italy, and she has been working on food chemistry and technology, food packaging hygiene, chemical and technological features of cheeses, study of sensorial features with reference to HACCP researches, and mandatory food traceability. Dr. Micali is also a lecturer in different sectors, including professional training. Her published works include 'The maximum water absorption in cheeses. Tripartite networks of absorbed molecules per nitrogen' (2009), The Chemistry of Thermal Food Processing Procedures (2016), and 'Traceability in the Cheesemaking Field. The Regulatory Ambit and Practical Solutions' (2016).

Chapter 1
Food Sharing and the Regulatory Situation in Europe. An Introduction

Abstract The current food production in the industrialised world is apparently facing and interesting paradox: the intensive flow of many food and beverage commodities, year by year, on the one hand, and the concomitant elimination of unused portions of produced edible products. According to the Food and Agriculture Organization, 88 million tonnes of food produced for human consumption are get wasted annually in the European Union, and associated costs exceed 140 billion Euros. Apparently, the greatest responsibilities for food waste are reported in countries with medium/high-income values if compared with developing nations. However, the genesis of wasted food appears questioned and extremely debated: food waste appears after harvest and during processing steps in developing countries and at the retail and consumer level in industrialised nations. Certainly, food waste is a phenomenon occurring in industrialised countries such as Germany: a notable part of food waste happens because of consumers' behaviour, although food retailers may give a significant contribution. The remaining part appears to be localised in the food industry. Consequently, the recently observed 'food sharing' communities have been created with a basic aim: to save food and give it to suffering people 'for free'. Could food sharing may be a solution? Organisations such as food banks and social supermarkets should be analysed and evaluated; specific regulations could be elaborated. Otherwise, recovered (and possibly degraded) products could be unsafe. This chapter explores the current situation and the regulatory definition of food sharing, with specific relation to different European countries.

Keywords European Union · Food bank · Food sharing · Food supply chain · Food waste · Social supermarket · Sustainable Development Goals

Abbreviations

BOGOF	Buy-one-get-one-free
EU	European Union
FAO	Food and Agriculture Organization of the United Nations
F&B	Food and beverage

© The Author(s), under exclusive license to Springer Nature Switzerland AG 2019
A. Pellerito et al., *Food Sharing*, Chemistry of Foods,
https://doi.org/10.1007/978-3-030-27664-5_1

FSC   Food supply chain
FW   Food waste
FFV   Fresh fruits and vegetable
SOMA   Soma-Sozialmarkt
SSM   Social supermarket
SDG   Sustainable Development Goal
USA   United States of America
UK   United Kingdom
WRAP   Waste and Resources Action Programme

1.1 An Introduction to the New Phenomenon of Food Sharing in Europe

The current food production in the industrialised world is apparently facing and interesting paradox: the intensive flow of many food and beverage (F&B) commodities, year by year, on the one hand, and the concomitant elimination of notable and unused portions of the globally produced F&B items. This phenomenon is easily observed today in the industrialised world, and the European Union (EU) is not an exception [1, 2].

According to the Food and Agriculture Organization (FAO), 88 million tonnes of food produced for human consumption are get wasted annually in the EU, with associated costs approximately estimated around 143 billion Euros [3–6]. Apparently, the greatest responsibilities for food waste are reported in countries with medium- and high-income values, with a level of wasted food approximately equal to 95–115 kg per person each year. On the contrary, developing countries waste just 6–11 kg of edible food per person each year, as shown in Fig. 1.1 [7].

However, the genesis of wasted food appears questioned and extremely debated at present. In this ambit, several observers claim that food loss and waste are quantitatively the same amount in industrialised and developing countries, while some difference should be mentioned when speaking of 'waste-generating' steps in the ambit of the food production and consumption life cycle. In other words, it has been reported that the main portion of food wastes in developing countries (over 40%) happens after harvest and during processing steps. On the other side, food wastes seem more abundant and quantitatively notable at the retail and consumer level in industrialised countries. Once more, the amount of 'notable' or 'important' food waste is $\geq 40\%$ in the last situation: this level is considered an alarming bell when speaking of abnormal exit of food and beverage commodities out of the normal food production and consumption life cycle [8].

In detail, and with reference to developing countries only, food waste (FW) appears limited to the first food supply chain (FSC) stages only. The first step in the FSC is the production of sustenance from the agricultural sector where both farming and cultivation generate waste or sub-products: organic waste (i.e. cornstalk), nourishment

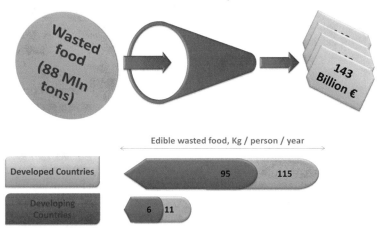

Fig. 1.1 Current situation in the European Union with relation to wasted foods, their economic value, and observed differences when speaking of 'wasting' behaviours in countries with medium/high-income and developing nations [3–7]

waste or sustenance misfortune (i.e. low-quality fruits or vegetable, imperfect products left in the field, good products or co-products with a low or absent commercial value).

Harm amid transport or non-fitting transport frameworks, issues amid capacity, misfortunes amid preparing or sullying, and/or wrong bundling are the main reasons that lead to the production nourishment misfortunes and nourishment squander all through the whole creation stage production phase.

Food losses happen for the most part at the beginning of the FSC. What about causes? Apparently, food losses should be notable and frequently observed when restricted specialised or budgetary assets in agriculture, reaping, gathering, or transporting nourishment can be assessed. On the other hand, this phenomenon is differently named in developed countries (and European member states belong to this category): food losses are somewhat called 'food waste' and happen mainly towards the end of the FSC. The interested steps are called 'appropriation', 'retail', and 'last utilisation' (by the final consumer): wrong practices like overbuying sustenance for customers, overstocking, or esthetical norms at the retail level are often reported [9].

At present, the European Commission is trying to find adequate solutions in order to significantly reduce food waste. In other words, should this assignment be reached, the EU would easily save money and lower the ecological effect of nourishment creation and utilisation (environmental impact) correlated with food production and consumption. Indeed, squandered sustenance favours overuse of water and petroleum derivatives, with the additional expansion of ozone-harming substance emanations, i.e. methane and carbon dioxide emerging from the debasement of sustenance in landfills [10].

The term 'sharing' often alludes to two distinct meanings. Regularly, it indicates anything that is shared between or among at least two individuals. In addition, it can suggest that two individuals at least are characterised by some mutual traits [11].

On a general level, sharing cannot be considered a new phenomenon. On the contrary, it has presumably been the most fundamental type of economical distribution in human societies for a hundred thousand years. Besides being perceived as a regular inter-family habit, the sharing of nourishment among various family units has been initially depicted by researchers based on primitive and contemporary hunter-gatherer communities [12].

At present, one of the biggest challenges worldwide is to ensure the global population with the indispensable food amount per capita, while around one billion people suffer globally hunger [13]. In the not distant future, an increment in the global population and climate changes will influence food production and food availability, with other correlated modifications such as climate change. For these reasons, feeding the total population will turn out to be even more difficult [14, 15].

The reliable assurance of food safety has been one of the most urgent problems to solve on the political agenda in the last years because of different factors such as unpredictable nourishment costs, the utilisation of farming yield as biofuel or to feed animals, and the occurrence of dry seasons [16]. Some provisions have been suggested in order to deal with growing difficulties when speaking of:

(a) Feeding opportunities for the total population, and
(b) The increase in food security without threading the environment (for example, stopping the development of farms, in particular in the equator area; improving efficiency during the cropping phase; changing diet habits; and diminishing waste).

Though the application of above-mentioned measures, the amount of food production could be doubled if available resources are used without increasing ecological effects [17]. Recently, the European Project 'FUSIONS' defined FW as 'any food, and inedible parts of food, removed from the food supply chain to be recovered or disposed (including composting, anaerobic digestion, bio-energy production, co-generation, incineration, disposal to sewer, landfill, or discarded to sea)' [3]. Figure 1.2 shows a simplified description of the FW definition. A dedicated analysis of food waste 'occasions' can demonstrate that it is reasonable to ascertain the total nourishment losses crosswise over various FSC steps in industrialised countries, where there is the chance to collect waste data regarding food waste. In the UK, the extent of sustenance and drink squander is roughly 14 million tonnes, of which 20% is apparently caused by nourishment processing, delivery, and trade processes. Family units are responsible for the biggest single contribution in FW terms. On the other side, precise and consistent evaluations of other post-consumer wastes (hospitality and institutional sources) are still not provided. The estimated total waste per annum originated from nourishment and drink manufacturing and processing is 5 million tonnes, where around 2.6 million tonnes is figured to be sustenance squander. Additional 2.2 million tonnes are related to by-products which the final destination is modified for (from human food to animal feed) [18].

Fig. 1.2 According to the European Project 'FUSIONS', food waste (FW) has a single 'exit' point from the FSC and different recovery or disposal options

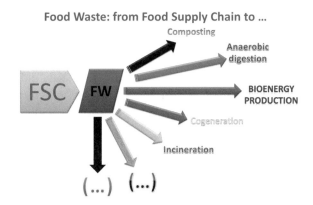

Surveys concerning FW production have established that a significant FW amount comes from sectors concerning perishable products such as meat and poultry, fresh fruits, and vegetables (FFV). This FW portion is represented for the most part by by-products and premade food products that remain unsold [19–23].

In the UK, the market power is under the control of a few vast retailers which practice their control over 7000 providers inside the area. In order to avoid cancellations from the dedicated list, food manufacturers prefer to overproduce in case they are asked to deliver extra quantities at short notice. For manufacturers of grocery stores' own brands, overproduction enclosed in a packet must be sold *in loco* without different options (e.g. to sell foods elsewhere). In case of unsold products, it ends up plainly squander; nonetheless, the sector is able to reuse the greater part of nourishment waste produced [24, 25]. In detail, family unit nourishment squander can be defined as sources of sustenance and beverages that are expended inside the home; retail, home-grown food, and takeaways are included in this definition.

Disposal routes are considered as domestic waste streams. Must be taken into account that this does not include noteworthy amounts of sustenance and beverage eaten 'on-the-go', during the work hours or served by catering organisations [25].

1.2 Food Waste... Why? A Social Analysis Across the Globe

Many studies have tried to identify which kinds of goods mainly get wasted. It has been reported that the most significant amount is represented by the freshest nourishment items that are considered the uppermost part of food waste. FFV are frequently in the list of the most-squandered items, trailed by different perishables like pastries, items produced by milk transformation, meat, and fish [26–28].

A significant variety in wastage rates has been frequently observed for various nourishment sorts: according to the Waste and Resources Action Programme (WRAP), 7% of milk purchases, 36% of bakery, and over 50% of lettuce/leafy salads (by weight) are wasted, while Jones and co-workers recorded comparable fluctuations in the typical wastage rates for different goods items [25, 29].

Nourishment and drink classifications are not completely reliable crosswise over investigations. However, Thonissen [28] noticed an uncommon large amount of food waste made by dairy products in the Netherlands, differently from Turkish data; the highest proportion of food waste was represented by wasted FFV [26].

Some factors, which may explain the fluctuation in the amount of household-produced FW, have been identified:

(a) Household size and composition
(b) Household income
(c) Household demographics
(d) Household culture.

With reference to household composition and related size, some studies from the UK [25, 30, 31] and the USA [32] demonstrate that nourishment wastage was altogether affected by the family composition, with grown-ups squandering more compared with youngsters, and bigger families squandering less per individual than littler families. In addition, family units composed by single person tend to discard more food per capita, and family units with kids tend to squander more than families without kids, despite the fact that rates fluctuate with the youngsters' age.

With reference to household income, the greater part of studies suggests that family units with low-income rates produce less waste of food compared with the family with higher income [31, 33].

With concern to household demographics, some research in the UK [31, 33] and in Australia [34] seem to suggest a bigger FW production by young people, who trash away more food than older people, with retired households trashing away minimum quantities (such family units generally consist of relatively fewer people).

Finally, the importance of household culture should be considered. There are some evidences that culture plays a role in the production of sustenance wastage.

1.3   Food Waste… As the Opposite of Food Sharing

The two main reasons which lead to unnecessary food waste could be mentioned as follows:

(a) 'Goods are not consumed before they get rotten', and
(b) 'An excessive amount of food is cooked, prepared, or served'.

The first of these statements should be plain enough without the need of further explanations (generally, it is considered as the historical 'legacy' of previous war or famine times) [27]. With relation to this peculiar declaration, food is not utilised in time. This spreads sustenance and beverage waste since it passed a date mark (e.g. a 'utilisation by' or 'best earlier' date), has gone mildew covered or lost its organoleptic properties [35].

The following cases belong to this class:

(a) Not eating goods needing to be consumed before than others, as consumers prefer eating what is more appealing to them rather than what they have already available at home;
(b) Not planning a regular cleaning of cabinets, coolers and coolers keeping in order to consume first old or forgotten food or dispose of undesirable sustenance items;
(c) Paying high attention to food hygienic conditions and 'best-before' dates on food labelling;
(d) Disappointment due to the taste of the nourishment—particularly sustenance left by youngsters.

On the other side, the second declaration should be analysed carefully.

In the larger part of cases, an excessive amount of food may be cooked at home, yet covering additionally situations where nourishment was harmed amid handling (e.g. consuming nourishment). This group of similar behaviours could be alluded to as 'leftovers'. The following cases belong to this class:

(1) Shopping an excessive amount of goods by and large, in particular in case of special offers. One of the most known situations is mentioned and recognised with the acronym 'buy-one-get-one-free' (BOGOF) [36–40];
(2) Purchasing an excessive amount of fresh (e.g. fruit and vegetables) to accomplish a more beneficial eating style and to attempt food experimentation;
(3) Preparing an excessive amount of food in general.

Another way of research and discussion with relation to FW causes should consider the active role of consumers. Food purchasers can be divided into three general classes in a pure FW discussion [35]:

(1) High food wasters—they trough away 'a considerable amount', 'a sensible sum', or 'a few' uneaten sustenance squander (30%).
(2) Medium food wasters—meaning they trough away 'a bit' uneaten nourishment squander (27%).
(3) Low food wasters, claiming they trough away 'scarcely any' or 'none'—(43%).

More in general, customers can be generally partitioned into two parts, with 50% of them responsive to the problem, while the remaining 50%—in any event at the present time—shows up lack of interest for the issue.

These groups can be further split roughly in the following proportions:

(a) Around 13% of purchasers give off an impression of being responsive to and effectively connected with the matter. They declare to be exceptionally annoyed by nourishment squander and try their best to decrease the total amount of food they trash away
(b) Around 28% of food consumers give off an impression of being open, however, aloof, in that they are 'genuinely' annoyed or potentially go to a 'considerable lot' of push to diminish nourishment squander. Their worry, along these lines, is less strong and less prone to convert into behavioural reactions

(c) An approximate amount of food consumers (26%) show up not to be especially open to sustenance squander, and they claim they are just to a small extent annoyed by the idea. They seem disengaged instead of effectively opposed to

(d) Finally, the remaining 33% of food purchasers seem both unworried regarding this topic and, few of them, impervious to endeavours that could support a lessening in the discarded sum.

Among the individuals who are bothered, three reasons to a great extent represent purchasers' worries: a misuse of cash, a feeling of squandering 'great sustenance', and a generalised feeling of blame. The health of the planet is not considered to be among the main problems, rather secondary and optional concern. These ideas and their consequences are amazingly steady over the populace all in all. Just among the target, socio-demographic groups are there some fascinating differences: social tenants, for instance, worry in a less extent for squandering 'great' sustenance and are more prone to refer to the monetary cost [35].

Apparently, food consumers have not obtained a significant association between nourishment waste and ecological effect so far. Undoubtedly, purchasers energetically trust that waste coming from plastic, paper, and other parts of the packages is a more noteworthy natural concern compared with nourishment squander, while there is a boundless conviction that sustenance squander has no ecological effect at all since it is biodegradable. In addition, high nourishment wasters will probably be more youthful (under 45 years of age), coming from a lower social layer, living in private or social rented accommodation, working or being at home because of maternity or paternity, living in larger family units and with children [35].

The third analysis concerning FW should take into account environmental sustainability. Sustenance waste ought to be recovered or discarded without damaging people or the environment. Especially, it should not:

(a) Represent a threat for natural resources as water, air, soil, or to fauna and flora

(b) Cause problems trough noise or odours

(c) Represent a threat to landscape or areas of special importance, protected in accordance with the current legislation.

In order to comply with the above rules, waste management should occur according to a well-known policy: the 'reduce, reuse, recycle, and recover' approach. In addition, waste disposal should be considered [41].

The degradation of 'biodegradable waste' coming from household kitchens, eateries, catering, and retail premises and comparing leftovers from food industry can be responsible for a significant environmental impact due to odour emissions, release of noxious gas, like methane, into the atmosphere, leachate into the soil, and resulting increments in relative restoration costs [41].

Along these lines, the natural effect of sustenance squander is twofold. From one perspective, it is related to the exhaustion of regular assets utilised for its creation (e.g. soil consumption) and conveyance; then again, it identifies with the expenses related to the discarding process. Besides these natural expenses, must be taken into account significant financial expenses for makers, purchasers and organisations, and various social expenses.

As a consequence, the decrease in food waste on a global scale can represent therefore a crucial step to contrast climate change and environmental damages.

1.4 Food Waste and Regulatory Countermeasures in Europe

For the above-mentioned reasons, a crucial session of the Commission's new Circular Economy Package is the food waste tackle, with the aim to stimulate Europe's transition in the direction of a circular economy which will sponsor competitiveness in a worldwide scale, foster sustainable growth and finally will provide new job vacancies. One of the main purposes of the EU and the member states is to satisfy the Sustainable Development Goals (SDG), established on September 2015, as to reduce up to the half the nourishment waste for each person at the sellers and buyers level within 2030, and lessen nourishment losses during the food production and supply chains [42].

In order to support aims of the SDG targets for food waste reduction in the EU, the Commission will, in combination with other aspects of the strategy, encourage free food giveaway and the recuperation of previous foodstuffs and by-products deriving from food chain, which will be transformed in animal feed, or invested for anaerobic digestion, composting, and associated means, without compromising neither food nor feed safety. Besides, it will explore new ways to improve the use of data made available to the consumers, in particular, 'best-before' labelling.

In consideration of the key role of sharing economy in quickening sustainable consumption, Cohen and Munoz provided an integrated model to conjectured five perfect sharing classes: energy, merchandise, sustenance, mobility and transportation, and space sharing [43]. Especially, the sustenance division is for sure perceived as a strategical zone for sustainable consumption. On a basic level and both from a full-scale and small-scale viewpoint, nourishment sharing may positively affect every one of the three aspects of sustainable development by encouraging saving, supporting the creation and/or affirmation of existing social relations, and by diminishing waste production.

In September 2016, France has turned into the principal country on the planet to forbid markets from discarding or devastating unsold sustenance, driving them rather than giving it to foundations and food banks. According to a law passed unanimity by the French Senate, extensive shops will never again waste eatable sustenance moving towards its best-before date. Foundations will have the capacity to give out millions more of the free dinners every year to individuals attempting to access to food [44–46].

The main foundations, which will benefit from this law coming into effect, are no-profit organisations are 'food banks' and 'social supermarkets'. As discussed in the following sections, one main feature seems to concern these organisations in spite of their differences: both food sharing forms tend to consider 'fit for use' for

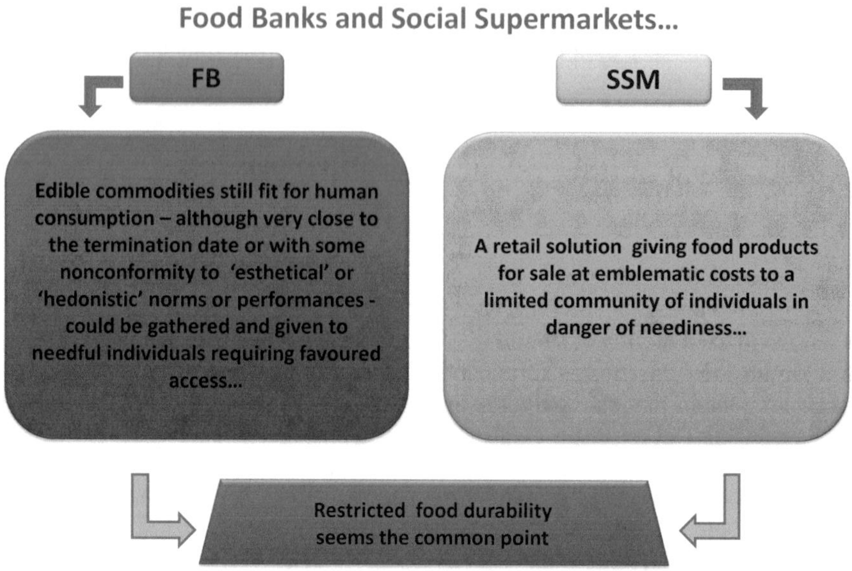

Food Banks and Social Supermarkets…

FB

SSM

Edible commodities still fit for human consumption – although very close to the termination date or with some nonconformity to 'esthetical' or 'hedonistic' norms or performances - could be gathered and given to needful individuals requiring favoured access…

A retail solution giving food products for sale at emblematic costs to a limited community of individuals in danger of neediness…

Restricted food durability seems the common point

Fig. 1.3 No-profit organizations such as 'food banks' and 'social supermarkets' generally consider 'fit for use' for human consumption all edible commodities even if these foods and beverages may be very close to expiration (or 'last before date') terms, or with some nonconformity to 'esthetical' or 'hedonistic' norms or intended performances

human consumption all edible commodities although very close to expiration (or 'last before date') terms, or with some nonconformity to 'esthetical' or 'hedonistic' norms or intended performances (Fig. 1.3). Consequently, an interesting research point may concern food durability (Chap. 4).

1.4.1 Food Banks

Food banks are a moderately new reality in Europe, in spite of the fact that they have for quite some time been a component of the North American welfare situation [19]. Despite worldwide financial grimness, the number of food banks is expanding; at present, there are entrenched models everywhere throughout the globe. Food banks can be classified as associations procuring nourishment, a lot of which would somehow or another be squandered, from farms, producers, wholesalers, retail locations, shoppers, and different sources, making it accessible to needful people through a system of group agencies.

Edible commodities still fit for human consumption—although very close to the termination date or with some nonconformity to 'esthetical' or 'hedonistic' norms or performances—could be gathered and given to needful individuals requiring favoured access [47].

Food banks are no-profit associations that gather sustenance from retailers and redistribute it to poor people for free. They represent the joining link that gathers donations in kind and money mostly from associations, yet even from people and governments; handle and store them; convey these donations to no-profit associations that redistribute the goods to or provide suppers for supporting poor people [48–51].

Because of this process, the balance between overproduction of food and food access—a problematic circumstance that very often is faced by the lowest layers of the populace in developed countries—has been partially restored.

Besides in France with the 'Banque Alimentaire de la Loire', food banks are as for now very active in some other nations such as Italy and Spain [51, 52]. With specific relation to Italy, the following banks—the 'Banco Alimentare' and other non-governmental organisations or retail initiatives such as 'Buon Fine'[1]—should be mentioned because of their role when speaking of gathering and donating unsold and still eatable goods to no-profit organisations [50, 53]. One of the Spanish banks is the 'Banco de Alimentos' [54, 55].

Web platforms and correlated mobile apps are becoming an integral part of these organisations. Banco Alimentare, for instance, provides a dynamic web platform and has launched the 'Bring the food' app service with the aim of making the project more accessible to needful people.

With relation to the UK, the food sharing project has been launched in 2016 by Tesco with the aim of giving unsold items to charitable associations. This project is incorporated into 'FareShare FoodCloud', a web-based service and into the 'Fare-Share Go' mobile app which connects grocery store and no-profit associations with the specific goal to share surplus nourishment.[2] Other alliances have been finalised between rival retailers belonging to different countries such as Tesco (UK) and Carrefour (France) [56, 57].

1.4.2 Social Supermarkets

Social supermarkets (SSM) are a retail solution where the retail service gets extra sustenance and purchaser products from organisations (e.g. makers, retailer) 'free-of-charge' putting them up for sales at emblematic costs to a limited community of individuals in danger of neediness.

SSMs' principal traits are [58]:

(1) A restricted variety of nourishment and household products;
(2) The distribution is restricted to documented individuals in danger of neediness.
(3) Shelf prices are brought down from 30% up to 70% if compared to conventional stores.

---

[1]Fostered by the retailer Coop in Italy.
[2]The service and the related project can be found at https://fareshare.org.uk/.

Soma-Sozialmarkt (SOMA), in Austria, is an example of social supermarkets that coordinates the entire retail process (from goods suppliers to point of sales and distribution) [59, 60]. SSMs are charitable associations; the incomes are utilised to cover costs (workforce, lease, transport, and so on) [58]. Other SSM examples are 'The Community Shop' (UK) and 'Dough Rauch' (USA).

A similar business model is 'the Daily Table' (USA), another sort of retail supermarket providing fresh food and basic need grocery and also prepared suppers at moderate costs. The Daily Table uses leftovers goods from general stores, producers, and nourishment merchants that would some way or another have been squandered.[3]

The ongoing digital innovation and related development have encouraged the advancement of web platforms and portable applications, considered an apparatus for helping SSM associations for the appropriation of nourishment, or favoured the start-up of new online SSM. The law abides by a grassroots campaign in France by customers, anti-poverty campaigners and those against the nourishment squander.

In April 2015, the French government officials released a list proposal for a national strategy against nourishment squander, raising an arrangement of ideas for avoidance, recuperation, and reusing [61].

Among these proposals, the following points should be mentioned:

(a) Create new ways of advertisement such as 'I don't toss things out anymore'. New types of communication on food waste reduction should be planned in order to reach people at home and in the working environment. An increased awareness would be accomplished with another national office responsible for squander aversion. A committed website would be set up and connected to interpersonal organisations, underlining constructive (as opposed to blaming based) informing.

(b) Clarify expiration dates on nourishment items. As most shoppers do not completely comprehend diverse sorts of termination dates now showing up on sustenance bundles, the wording should be made clearer, eminently using 'best earlier dialect' rather than 'consume preferably before'.

(c) Organise local food recovery days. A national, yearly 'food recovery day' would be set up to activate groups and organisations at the local level, including supermarkets and eateries. In this way, organisations and people would have the chance to give shelf-stable or fresh products and for nearby associations to enrol potential volunteers and assemble organisations.

(d) Ban supermarkets from throwing away edible excess food. Retailers should have a legal commitment to diminish, reuse, or reuse their additional sustenance, with a monetary penalty in the case that they do not observe it, according to current laws. For example, grocery stores will be urged to make 'zero-waste areas' where they offer items near their lapse dates and to assign 'zero-waste trainers' to bring issues to light among workers and help food achieving the termination of its saleable life. All unsold items should be donated by supermarkets for free when they are eatable. Otherwise, natural waste (expired food) should be reused for other purposes.

---

[3]This service can be found at https://dailytable.org/snap/.

(e) Order donations to beneficent associations. Markets will be obliged to make arrangements with approved no-profit associations so as to give additional sustenance that the association demands. The accord is meant to guarantee the quality and convenience of donations. Since assess motivating forces are generally high in France—60% of nourishment esteem and associated coordination expenses—they will be conceded just for items which have not reached their last utilisation date and can really be utilised (i.e. given over one day before the lapse date). The law made by the French Parliament in July 2015 states that markets bigger than 400 m^2 should set up concurrences with approved charitable associations.

(f) Ban destruction of edible food. It is a normal habit for shops engaged in handling food to drop some bleach on wasted goods to ensure it will not be collected and reused. The bill, approved by the Parliament on July 2015, forbids stores to decimate still eatable groceries. State controllers will screen this to guarantee that items are crushed just in instances of genuine nourishment danger.

(g) Allow the donation of unwanted 'house' brand products. Nourishment producers will now have the capacity to give things that are dismissed by a retailer, regardless of the possibility that they show a retailer's own 'house' mark. (To-date, makers have normally been punished by retailers from giving these items, in order to ensure the retailers' image picture).

(h) Adjust portion and packaging size. A lot of shopper sustenance squander is driven by excessive dishes sizes and too big bundle sizes for basic supply items. In spite of the fact that directions would not be set up, the legislature would promote good practices in the retail and food service sectors, for example, more properly estimated dishes, partition sizes, and bundling choices; the chance to pay according to the weight of the products and other innovations. The administration itself tries to settle down a positive example regarding waste prevention and lessening. For occurrence, some state-funded schools are as of now revaluating the amount of food that should be served for each meal

(i) Clarify the 'last before date' meaning. Worries have emerged that last before date could be misused by retailers in an illegal way with their providers, for example, to refuse items if they have in over number. Accordingly, the suggested arrangement should regulate the way in which dates are utilised as a part of supply contracts. Additional, the law, approved by the Parliament on July 2015, in accordance with European controls, forbids termination dates on items, for example, crisp delivery, wine, bread kitchen merchandise, vinegar, salt, sugar, and sweets. Also, France will push for a more extended rundown of items that are not in need of requiring 'best earlier' dates at the European level.

(j) Encourage reuse of food by-products for animal feed. Numerous items not suitable for human utilisation can be recycled to feed animals. So, dispose of sustenance utilised to nourish animals must be harmless, nutritious, cheap, and dependable on a solid premise to address the issues of animal makers. According to this line, 'by-products' (that could be utilised as animal feed) must be separated from 'biowaste'. European directions would likewise need to advance keeping in mind the end goal to better characterise 'by-products' and support their utilisation as animal feed.

1.5 Food Sharing Versus Food Waste

A different strategy for facing the FW problem is the 'food sharing' approach.

The sharing food trend is nowadays achieving again popularity. Nevertheless, this practice is not modern, rather an ancient habit, finding its cultural and social roots in ancient times. The reason for the recovery of this social phenomenon is mostly related to food overproduction [61].

Food is graspable, tangible, visible, and related to living organisms. Everybody needs to eat edible and well-stored food to survive; as a result, the principal target of the food sharing community is to avoid the waste of food by gifting food commodities to other persons or institutions [62]. Indeed, should some food be wasted along the FSC, this waste would not concern just the good itself rather also the energy and efforts that have been invested in growing, nurturing, harvesting, producing, packaging, and transporting activities.

In industrialised countries such as Germany, the principal FW source is identified with household consumers and retailers. These subjects cause up to 40% of all FW. The remaining 60% is due to processes as production, agriculture, post-harvesting, and processing activities [7].

The idea behind the food sharing community is in line with the creation of sustainable food behaviour [63]. In particular, it calls for a reduction in the consumption; the conservation of the resources; and the creation of new types of socio-economic relationships.

The most common models of food sharing—in particular, distribution points, in which sustenance is swapped—are gift-based, often with the goal of turning nourishment squander, surpluses and losses into new values. Since sustenance is needed worldwide, the goal is to reduce overconsumption, nourishment wastage, and to give at last supportable eating resolutions [64].

The accomplishment of the sharing economy essentially relies upon the presence of network platforms, empowered by information and communication technologies capable to interface customers' needs to the sharing economy activities. Subsequently, various economical activities in the light of new sorts of utilisation models are increasing their power in the most part of developed countries. Different business models, called sharing economy exercises, are developing in numerous fields going from transportation (e.g. Car2go, Uber, convenience (e.g. Airbnb), to funds (e.g. Indiego) [65]). In this way, the improvements of Internet technologies, on the one hand, and the collapse of the consumerist society behaviour, on the other hand, have brought to the development and dispersion of this new economic model. Currently, buyers do not pay their attention just to possession: their interest is focused on goods and service [66]. Sharing and community-oriented practices are portrayed by brief access to merchandise and services and reliance on Internet channels.

Nowadays, sharing models exist at all stages of the supply chain, from the first step of the production up to the last step of the distribution. Examples include the following cases:

(a) Platforms, such as Landshare (www.landshare.net) that connect individuals who share the same interest for home-grown food, allow them to gather together and connect those who are in surplus of land to share with people who do not have enough land or not land at all for cultivating food [67].
(b) Pop-up or temporary restaurant platforms, such as Grub Club (http://grubclub.com), which creates a network of food lovers and gourmet chefs in temporary home/restaurant settings [68];
(c) Supper clubs and meal-sharing platforms, such as Casserole Club (www.casseroleclub.com) with the goal of lessening increasing social problems of isolation and undernourishment among older social layers and helping the creation of social networks between people and their neighbours [69].

Other sharing examples are:

(1) I Food Share (www.ifoodshare.org), a web platform in which is possible to offer free unused food [70];
(2) Viseat (https://viseat.com), a social platform by which users can arrange meals and other culinary events in their houses [71];
(3) BonAppetour (http://bonappetour.com), a social dining marketplace that connects tourists with local hosts to let them experience the taste of home-made meals, besides dinner parties, cooking classes, etc. [1].

Another good example in which people who are willing to offer or receive food for free (in order to avoid its waste), and can be connected through the web and social technologies, is the German community foodsharing.de (https://foodsharing.de) [72]. This and similar platforms are able to offer and accept food without money transactions.

foodsharing.de originated in Cologne when several committed people came together to form an association. The ideological father of this movement is Valentin Thurn, a documentary film-maker. Thurn had created a documentary called 'Taste the Waste' in which the problem of food being wasted is presented from different perspectives including farmers, wholesalers, food retailers, and consumers [73]. Through the use of social networks and platforms, the amount of this waste can be reduced significantly. An example of a very well-known social platform is Facebook, which provides the possibilities to connect people locally and globally.

Such a social mean can be adopted in restrict local geographical areas to build up interpersonal relationships, organise social events and services, share information and opinions, with the final result of an interaction that happens online as well offline.

For these and other reasons, food sharing approach is suitable in theory for all levels of the FSC. Other solutions such as the recycle of unused and still eatable foods may be good enough by the industrial viewpoint [72, 74–80], but could be also used as part of a more complex and multi-operational strategy with food sharing [81–91].

References

1. Ganglbauer E, Fitzpatrick G, Subasi Ö, Güldenpfennig F (2014) Think globally, act locally: a case study of a free food sharing community and social networking. In: Proceedings of the 17th ACM conference on Computer supported cooperative work & social computing (CSCW), Baltimore, MD, 15–19 Feb 2014. https://doi.org/10.1145/2531602.2531664
2. Michelini L, Principato L, Iasevoli G, Grieco C (2016) The social value of the sharing economy: a classification of innovative models in the food industry. In: Proceedings of the XXVIII Convegno annuale di Sinergie 'Management in a Digital World: Decisions, Production, Communication', Udine, pp 95–110, 9–10 June 2016
3. Anonymous (2019) Food waste: definition. EU Fusions, Brussels. Available https://www.eu-fusions.org/index.php/about-food-waste/280-food-waste-definition. Accessed 27 May 2019
4. Slow Food (2019) Food waste. Slow food, Bra. Available https://www.slowfood.com/sloweurope/en/topics/food-waste/. Accessed 27 May 2019
5. FAO (2017) EU and FAO bring combined weight to bear on food waste, antimicrobial resistance. Food and Agriculture Organization of the United Nations (FAO), Rome. Available http://www.fao.org/news/story/it/item/1040628/icode/. Accessed 27 May 2019
6. Stenmarck Å, Jensen C, Quested T, Moates G (2016). Estimates of European food waste levels. Available http://www.eu-fusions.org/phocadownload/Publications/Estimates%20of%20European%20food%20waste%20levels.pdf. Accessed 27 May 2019
7. Godemann J, Michelsen G (eds) (2011) Sustainability communication: interdisciplinary perspectives and theoretical foundations. Springer, Dordrecht
8. Gustavsson J, Cederberg C, Sonesson U, van Otterdijk R, Meybeck A (2011) Global food losses and food waste—extent, causes and prevention. Food and Agriculture Organization of the United Nations (FAO), Rome. Available http://www.fao.org/3/a-i2697e.pdf. Accessed 27 May 2019
9. Griffin M, Sobal J, Lyson TA (2009) An analysis of a community food waste stream. Agric Hum Values 26(1):67–81
10. Hall KD, Guo J, Dore M, Chow CC (2009) The progressive increase of food waste in America and its environmental impact. PLoS ONE 4(11):e7940
11. Zvolska L (2015) Sustainability potentials of the sharing economy: the case of accommodation sharing platforms. Master thesis IIIEE, Lund University, Sweden
12. Hunt RC (2000) Forager food sharing economy: transfers and exchanges. Senri Ethnological Stud 53:7–25
13. Naylor RL (2011) Expanding the boundaries of agricultural development. Food Secur 3:233–251
14. van Beek CL, Meerburg BG, Schils RL, Verhagen J, Kuikman PJ (2010) Feeding the world's increasing population while limiting climate change impacts: linking N_2O and CH_4 emissions from agriculture to population growth. Environ Sci Policy 13(2):89–96. https://doi.org/10.1016/j.envsci.2009.11.001
15. Godfray HCJ, Beddington JR, Crute IR, Haddad L, Lawrence D, Muir JF, Pretty J, Robinson S, Thomas SM, Toulmin C (2010) Food security: the challenge of feeding 9 billion people. Science 327(5967):812–818. https://doi.org/10.1126/science.1185383

16. Donner SD (2007) Surf or turf: a shift from feed to food cultivation could reduce nutrient flux to the Gulf of Mexico. Glob Environ Change 17(1):105–113. https://doi.org/10.1016/j.gloenvcha. 2006.04.005

17. Kummu M, De Moel H, Porkka M, Siebert S, Varis O, Ward PJ (2012) Lost food, wasted resources: global food supply chain losses and their impacts on freshwater, cropland, and fertiliser use. Sci Total Environ 438:477–489. https://doi.org/10.1016/j.scitotenv.2012.08.092

18. Lee P, Peter W, Hollins O (2010) Waste arisings in the supply of food and drink to households in the UK. Waste and Resources Action Programme (WRAP), Banbury

19. Segrè A, Gaiani S (2012) Transforming food waste into a resource. Royal Society of Chemistry Publishing, London

20. Block LG, Keller PA, Vallen B, Williamson S, Birau MM, Grinstein A, Haws KL, LaBarge MC, Lamberton C, Moore ES, Moscato EM, Walker Reczek R, Tangari AH (2016) The squander sequence: understanding food waste at each stage of the consumer decision-making process. J Pub Policy 35(2):292–304. https://doi.org/10.1509/jppm.15.132

21. Youngs AJ, Nobis G, Town P (1983) Food waste from hotels and restaurants in the UK. Waste Manag Res 1(4):295–308

22. Barilla (2019) Food waste: causes, impacts and proposals. Barilla Center for Food & Nutrition, Parma. Available https://www.barillacfn.com/m/publications/food-waste-causes-impact-proposals.pdf. Accessed 27 May 2019

23. STREFOWA (2019) Situation of food waste in Italy. Strategies to Reduce Food Waste in Central Europe (STREFOWA), Interreg CENTRAL EUROPE Programme, Wien. Available http://www.reducefoodwaste.eu/situation-on-food-waste-in-italy.html. Accessed 27 May 2019

24. C-Tech Innovation Ltd (2004) United Kingdom food and drink processing mass balance. A Biffaward Programme on Sustainable Resource Use, Chester

25. Quested T, Johnson H (2009) Household food and drink waste in the UK. Waste and Resources Action Programme (WRAP), Banbury

26. Pekcan G, Köksal E, Küçükerdönmez Ö, Özel H (2006) Household food wastage in Turkey. Food and Agriculture Organization of the United Nations (FAO), Rome

27. Mosby I, Galloway T (2017) 'The abiding condition was hunger': assessing the long-term biological and health effects of malnutrition and hunger in Canada's residential schools. Brit J Can Stud 30(2):147–162. https://doi.org/10.3828/bjcs.2017.9

28. Thönissen R (2009) Food waste: The Netherlands. Presentation to the EU Presidency Climate Smart Food Conference, Lund, Sweden, Nov 2009

29. Parfitt J, Barthel M, Macnaughton S (2010) Food waste within food supply chains: quantification and potential for change to 2050. Philos Trans R Soc Lond B Biol Sci 365(1554):3065–3081. https://doi.org/10.1098/rstb.2010.0126

30. Wenlock R, Buss D (1977) Wastage of edible food in the home: a preliminary study. J Hum Nutr 31(6):405–411

31. Osner R (1982) Food wastage. Nutr Food Sci 82(4):13–16. https://doi.org/10.1108/eb058904

32. Van Garde SJ, Woodburn MJ (1987) Food discard practices of householders. J Am Diet Assoc 87(3):322–329

33. Lyndhurst B, Cox J, Downing P (2007) Food behaviour consumer research: quantitative phase. Waste and Resources Action Programme (WRAP), Banbury. Available http://www.wrap.org. uk/sites/files/wrap/Food%20behaviour%20consumer%20research%20quantitative%20jun% 202007.pdf. Accessed 27 May 2019

34. Hamilton C, Denniss R, Baker D (2005) Wasteful consumption in Australia. Discussion Paper 77. The Australia Institute, Manuka, pp 2–46, Mar 2005

35. Press Association (2014) Buy-one-get-one-free offers 'should be scrapped to cut food waste'. The Guardian, https://www.theguardian.com. Available https://www.theguardian.com/ business/2014/apr/06/buy-one-get-one-free-food-waste-supermarkets. Accessed 27 May 2019

36. Anonymous (2018) From BOGOF offers to single-use packaging, expert panel discusses different food waste issues. Loughborough University, https://www.lboro.ac.uk. Available https://www.lboro.ac.uk/news-events/news/2018/november/expert-panel-discuss-food-waste-issues/. Accessed 27 May 2019

37. Anonymous (2018) Supermarket 'Bogof' deals criticised over food waste. British Broadcasting Corporation Ltd., (BBC), https://www.bbc.com. Available https://www.bbc.com/news/uk-26908613. Accessed 27 May 2019
38. Spiegel U, Benzion U, Shavit T (2011) Free product as a complement or substitute for a purchased product-does it matter. Modern Econ 2(2):124–131
39. Ramanathan U, Muyldermans L (2010) Identifying demand factors for promotional planning and forecasting: a case of a soft drink company in the UK. Int J Prod Econ 128(2):538–545. https://doi.org/10.1016/j.ijpe.2010.07.007
40. Morone P, Papendiek F, Tartiu VE (eds) (2017) Food waste reduction and valorisation: sustainability assessment and policy analysis. Springer, Cham. https://doi.org/10.1007/978-3-319-50088-1
41. European Commission (2019) EU actions against food waste. European Commission, Brussels. Available https://ec.europa.eu/food/safety/food_waste/eu_actions_en. Accessed 27 May 2019
42. Cohen B, Muñoz P (2016) Sharing cities and sustainable consumption and production: towards an integrated framework. J Clean Prod 134(1):87–97. https://doi.org/10.1016/j.jclepro.2015.07.133
43. Mourad M (2016) Recycling, recovering and preventing "food waste": competing solutions for food systems sustainability in the United States and France. J Clean Prod 126:461–477. https://doi.org/10.1016/j.jclepro.2016.03.084
44. Mourad M (2015). France moves toward a national policy against food waste. Dissertation, Natural Resources Defense Council, New York
45. Perchard E (2016) French food waste law passes unanimously. Resource Media Limited, Bristol. Available https://resource.co/article/french-food-waste-law-passes-unanimously-10826. Accessed 27 May 2019
46. Wells R, Caraher M (2014) UK print media coverage of the food bank phenomenon: from food welfare to food charity? Brit Food J 116(9):1426–1445. https://doi.org/10.1108/BFJ-03-2014-0123
47. Gentilini U (2013) Banking on food: the state of food banks in high-income countries. IDS Working Papers 415:1–18. https://doi.org/10.1111/j.2040-0209.2013.00415.x
48. Mohan S, Gopalakrishnan M, Mizzi PJ (2013) Improving the efficiency of a non-profit supply chain for the food insecure. Int J Prod Econ 143(2):248–255. https://doi.org/10.1016/j.ijpe.2011.05.019
49. Baglioni S, De Pieri B, Tallarico T (2017) Surplus food recovery and food aid: the pivotal role of non-profit organisations. Insights from Italy and Germany. Voluntas 28(5):2032–2052. https://doi.org/10.1007/s11266-016-9746-8
50. González-Torre PL, Coque J (2016) How is a food bank managed? Different profiles in Spain. Agric Hum Values 33(1):89–100. https://doi.org/10.1007/s10460-015-9595-x
51. Santini C, Cavicchi A (2014) The adaptive change of the Italian Food Bank foundation: a case study. Brit Food J 116(9):1446–1459. https://doi.org/10.1108/BFJ-06-2014-0201
52. Bandera L, Lodi Rizzini C, Maino F (2016) Povertà alimentare in Italia: le risposte del secondo welfare. Il Mulino, Bologna
53. Schneider F (2013) The evolution of food donation with respect to waste prevention. Waste Manag 33(3):755–763. https://doi.org/10.1016/j.wasman.2012.10.025
54. de Armiño KP (2014) Erosion of rights, uncritical solidarity and food banks in Spain. In: Riches G, Silvasti T (eds) First world hunger revisited. Palgrave Macmillan, London, pp 131–145. https://doi.org/10.1007/978-1-137-29873-7_10
55. Anonymous (2018) France's Carrefour and UK's Tesco sign own-brand food sharing deal. The Local Europe AB, Stockholm. Available https://www.thelocal.fr/20180806/frances-carrefour-and-uks-tesco-agree-on-food-sharing-alliance. Accessed 28 May 2019
56. Anonymous (2015) Tesco, un'app per il food sharing nel Regno Unito. Italiafruit News, Forlì. Available http://www.italiafruit.net/DettaglioNews/31821/mercati-e-imprese/tesco-unapp-per-il-food-sharing-nel-regno-unito. Accessed 28 May 2019
57. Holweg C, Lienbacher E, Zinn W (2010) Social Supermarkets-a new challenge in supply chain management and sustainability. Supply Chain Forum Int J 11(4):50–58. https://doi.org/10.1080/16258312.2010.11517246

58. Schneider F, Scherhaufer S, Montoux H, Gheoldus M, O'Connor C, Derain A (2015) Advancing social supermarkets across Europe. Available https://www.eu-fusions.org/phocadownload/feasibility-studies/Supermarkets/Advancing%20social%20supermarkets%20report.pdf. Accessed 28 May 2019

59. Garot G (2015) Lutte contre le gaspillage alimentaire: pro-positions pour une politique publique. Ministère de l'Agriculture, de l'Agroalimentaire et de la Forêt, Paris, 100p

60. Parisi S, Barone C, Sharma RK (2016) Chemistry and food safety in the EU. Springer International Publishing, Cham. https://doi.org/10.1007/978-3-319-33393-9

61. Evans D (2012) Binning, gifting and recovery: the conduits of disposal in household food consumption. Environ Plan D Soc Space 30(6):1123–1137. https://doi.org/10.1068/d22210

62. Heinrichs H (2013) Sharing economy: a potential new pathway to sustainability. Gaia 22(4):228–231

63. Davies AR, Doyle R (2015) Transforming household consumption: from backcasting to Home-Labs experiments. Ann Assoc Am Geogr 105(2):425–436. https://doi.org/10.1080/00045608.2014.1000948

64. Botsman R, Rogers R (2010) What's mine is yours: how collaborative consumption is changing the way we live, vol 5. Harper Collins Publishers, New York

65. Rifkin J (2000) The age of access. Penguin, Harmondsworth

66. Ikkala T, Lampinen A (2014) Defining the price of hospitality: networked hospitality exchange via Airbnb. In: Proceedings of the companion publication of the 17th ACM conference on computer supported cooperative work & social computing, ACM, Baltimore, MD, pp 173–176. https://doi.org/10.1145/2556420.2556506

67. Privitera D (2016) Describing the collaborative economy: forms of food sharing initiatives. In: Proceedings of the 2016 international conference on economic science for rural development, Jelgava, pp 92–98, 21–22 Apr 2016. Available http://llufb.llu.lv/conference/economic_science_rural/2016/Latvia_ESRD_43_2016-92-98.pdf. Accessed 28 May 2019

68. Marciante L (2014) Sharing in web society times: exchanging food. J Nutr Ecol Food Res 2(2):163–169. https://doi.org/10.1166/jnef.2014.1069

69. Albinsson PA, Yasanthi Perera B (2012) Alternative marketplaces in the 21st century: building community through sharing events. J Consum Behav 11(4):303–315. https://doi.org/10.1002/cb.1389

70. Bastasin G (2016) Social eating: una ricerca qualitativa su motivazioni, esperienze e valutazioni degli utenti della piattaforma Viseat.com. Dissertation, University of Pisa, Pisa

71. Mortara A, Fragapane S (2018) Vieni a mangiare da me? Un'analisi esplorativa del fenomeno del social eating. Sociol Comun 55(16):71–86. https://doi.org/10.3280/SC2018-055005

72. Brancoli P, Rousta K, Bolton K (2017) Life cycle assessment of supermarket food waste. Conserv Recycl 118:39–46. https://doi.org/10.1016/j.resconrec.2016.11.024

73. López CA, Butler BS (2013) Consequences of content diversity for online public spaces for local communities. In: Proceedings of the 2013 conference on computer supported cooperative work. ACM, San Antonio, pp 673–682, 23–27 Feb 2013. https://doi.org/10.1145/2441776.2441851

74. Minsaas J, Heie AC (1979) Recycling of do-mestic food waste. Conserv Recycl 3(3–4):427–438. https://doi.org/10.1016/0361-3658(79)90040-74

75. Shaw P, Smith M, Williams I (2018) On the prevention of avoidable food waste from domestic households. Recycling 3(2):24. https://doi.org/10.3390/recycling3020024

76. Scholz K, Eriksson M, Strid I (2015) Carbon foot-print of supermarket food waste. Resour Conserv Recycl 94:56–65. https://doi.org/10.1016/j.resconrec.2014.11.016

77. Refsgaard K, Magnussen K (2009) Household behaviour and attitudes with respect to recycling food waste–experiences from focus groups. J Environ Manag 90(2):760–771. https://doi.org/10.1016/j.jenvman.2008.01.018

78. Fehr M, Calcado MDR, Romao DC (2002) The basis of a policy for minimizing and recycling food waste. Environ Sci Policy 5(3):247–253. https://doi.org/10.1016/S1462-9011(02)00036-9

79. Da Cruz NF, Ferreira S, Cabral M, Simoes P, Marques RC (2014) Packaging waste recycling in Europe: is the industry paying for it? Waste Manag 34(2):298–308. https://doi.org/10.1016/j.wasman.2013.10.035

80. Comber R, Thieme A (2013) Designing beyond habit: opening space for improved recycling and food waste behaviors through processes of persuasion, social influence and aversive affect. Pers Ubiquit Comput 17(6):1197–1210. https://doi.org/10.1007/s00779-012-0587-1
81. Fait M, Scorrano P, Mastroleo G, Cillo V, Scuotto V (2019) A novel view on knowledge sharing in the agri-food sector. J Knowl Manag 23(5):953–974. https://doi.org/10.1108/jkm-09-2018-0572
82. Shirado H, Iosifidis G, Tassiulas L, Christakis NA (2019) Resource sharing in technologically defined social networks. Nat Comm 10(1):1079. https://doi.org/10.1038/s41467-019-08935-2
83. Sidali KL, Kastenholz E, Bianchi R (2015) Food tourism, niche markets and products in rural tourism: combining the intimacy model and the experience economy as a rural development strategy. J Sustain Tour 23(8–9):1179–1197. https://doi.org/10.1080/09669582.2013.836210
84. Marovelli B (2019) Cooking and eating together in London: food sharing initiatives as collective spaces of encounter. Geoforum 99:190–201. https://doi.org/10.1016/j.geoforum.2018.09.006
85. Hamari J, Sjöklint M, Ukkonen A (2016) The sharing economy: why people participate in collaborative consumption. J Assoc Inf Sci Technol 67(9):2047–2059. https://doi.org/10.1002/asi.23552
86. Pascucci S, Dentoni D, Lombardi A, Cembalo L (2016) Sharing values or sharing costs? Understanding consumer participation in alternative food networks. NJAS Wageningen J Life Sci 78:47–60. https://doi.org/10.1016/j.njas.2016.03.006
87. Tornaghi C (2017) Urban agriculture in the food-disabling city: (Re)defining urban food justice, reimagining a politics of empowerment. Antipod 49(3):781–801. https://doi.org/10.1111/anti.12291
88. Falcone PM, Imbert E (2017) Bringing a sharing economy approach into the food sector: the potential of food sharing for reducing food waste. In: Morone P, Papendiek F, Tartiu V (eds) Food waste reduction and valorisation. Springer International Publishing, Cham. https://doi.org/10.1007/978-3-319-50088-1_10
89. Kaushal LA (2018) The rise in the sharing economy: Indian perspective. In: Zakaria N, Kaushal LA (eds) Global entrepreneurship and new venture creation in the sharing economy. IGI Global, Hershey. https://doi.org/10.4018/978-1-5225-2835-7.ch007
90. Loh P, Agyeman J (2019) Urban food sharing and the emerging Boston food solidarity economy. Geoforum 99:213–222. https://doi.org/10.1016/j.geoforum.2018.08.017
91. Nishino N, Takenaka T, Takahashi H (2017) Manufacturer's strategy in a sharing economy. CIRP Ann 66(1):409–412. https://doi.org/10.1016/j.cirp.2017.04.004

Chapter 2
Food Sharing in Practice: The German Experience in Magdeburg

Abstract The sharing food trend is achieving again popularity at present in industrialised countries. The main reason seems correlated with food overproduction. 'Food sharing' communities have a basic aim: to avoid the waste of food by gifting food commodities to needful people or institutions. Indeed should some food be wasted along the food supply chain, this waste would also concern energy and efforts that have been invested in food-related activities. In industrialised countries such as Germany, the principal food waste source is identified with household consumers and retailers (up to 40% of all wasted foods), while the resting amount is reported to be caused during processes as production, agriculture, post-harvesting, and processing activities. In Germany, as indicated by a statistic from 2012, 11 million tonnes of food coming from food industries, trade, wholesale, and private households (60%) are wasted annually. The answer of German citizens is the sharing economy, and this project essentially relies upon the presence of physical distribution points and/or network platforms, empowered by information and communication technologies, with different applications. Nowadays, more than 200,000 'foodsavers' are engaged in this no-profit and non-commercial project in the Germany/Austria/Switzerland area. This Chapter described some of the local implementation efforts in Germany and a Magdeburg-based experience in particular.

Keywords Durability · Foodsaver · Food sharing · Food supply chain · Food waste · Nourishment · Sustainable Development Goal

Abbreviations

BMUB	Bundesministerium für Umwelt, Naturschutz, Bau und Reaktorsicherheit
EU	European Union
FSC	Food supply chain
FW	Food waste
WWII	Second World War

SDG   Sustainable Development Goal
USA   United States of America
UK   United Kingdom

2.1 Food Sharing in Germany

The sharing food trend is achieving again popularity at present (Sect. 1.5). The main reason seems correlated with food overproduction [1].

The main target of food sharing communities worldwide is to avoid the waste of food by gifting food commodities to other persons or institutions [2]. Indeed should some food be wasted along the food supply chain (FSC), this waste would also concern energy and efforts that have been invested in food-related activities. In industrialised countries such as Germany, the principal food waste (FW) source is identified with household consumers and retailers. These subjects cause up to 40% of all FW, while the resting amount is reported to be caused during processes as production, agriculture, post-harvesting, and processing activities [3].

The idea behind the food sharing community is in line with the creation of sustainable food behaviours [4]. In particular, it calls for reduction in the consumption; the conservation of the resources; and the creation of new types of socio-economic relationships.

The accomplishment of the sharing economy essentially relies upon the presence of physical distribution points and/or network platforms, empowered by information and communication technologies [5], with different applications including also transportation convenience, and funding opportunities [6, 7].

In Germany, as indicated by a statistic from 2012, 11 million tonnes of food coming from food industries, trade, wholesale, and private households are wasted annually. The main sources are private households, covering three-fifths of the food waste: every citizen of the German Federal Republic is estimated to through away 82 kg of nourishment a year. Surely, half of the food wasted in private households would be still in fully eatable conditions. Waste of broths, apple peels, or foodstuffs—which represents another fifth of the total FW amount—would be partly avoidable since it is only due to different consumption habits in the garbage. Only one-third of the total waste, which is composed for example of bones or banana peels, would not be avoidable. Only one-sixth of respondent citizens claimed that they have never thrown food in the garbage [8].

Almost three out of ten do it at least once a week. According to Metzler and co-workers, 1876 people from 21 years onwards were asked how often during their last month at home they have thrown food away [9]. Obtained and analysed results demonstrate that people who were in shortage of food during the Second World War (WWII) waste significantly less food compared with the later generations. One-third of the people who were born before 1945 never put drinks and food into the garbage. Another two-fifths of this (pre-) War generation throws a maximum once a month food away. Only every tenth disposed of at least once a week. These people tacitly demonstrate that FW has to be avoided: 'Goods are not consumed before they get rotten'. In general, this behaviour may be considered as the historical 'legacy' of previous war or famine times (Sect. 1.3).

In the post-WWII generation (generations from 1945 to 1954), this value is 16%. Nevertheless, the waste of eatable food even among the post-WWII generation is not common. One out of four never disposes food; another one-third maximum once a month.

Only one-sixth of the so-called baby boomers (generation from 1955 to 1964) never wastes food. Every fourth baby boomer throws food away at least once a week. Another quarter declares they waste food at least several times a month [10].

The Federal Government is involved in the implementation of Sustainable Development Goals (SDG), which were previously adopted by the United Nations in September 2015 [11]. The main of these SDG concerns Agenda 2030 targets for the reduction of preventable food waste.

In particular, the implementation of the objective SDG 12.3 is seen as a society-wide task, which can be successfully achieved only if all participants along the value-added chain collaborate, as solid commitments as could reasonably be expected, and advance social attention to the estimation of nourishment. It has been possible to highlight issues with the activity 'Too good for the ton!' (launched in March 2012) about the estimation of nourishment. Several background information and useful tips can be found on the website www.zugutfuerdietonne.de. Many different cooking ideas are available for the best remnants in the large recipe database. It is also possible to set personal recipe ideas. An interactive test can be used to find out the best way to lessen food waste. In the 'News' section, the initiative provides information on actions; in addition, experts can give advice and provide information on trade, politics, or science.

The difficulties of a reliable recording of food waste and losses in all areas of the value chain are shared by all European countries. For this reason, a methodological handbook has been created within the framework of the EU-FUSIONS project, which contains recommendations and a collection of possible procedures for the collection of food waste in Europe. It gives useful rules to the measurement of nourishment squander at various levels of the value chain, providing a common basis for gathering data and comparing them [12].

The possibility of a legal record of food waste is discussed in various ways in connection with the ongoing discussions of the so-called cycle management package within the framework of the revision of the European Waste Framework Directive. It has to be noted that such monitoring is possible only on the basis of standardised detection methods. However, recorded data concerning FW and related losses along the value chain can be taken by the Federal Government only on the base of voluntary agreements with interested stakeholders. Therefore, the German Federal Government depends on the cooperation of participants along the value chain. This action will also be the subject of the ongoing strategy process to develop action and research to lessen food losses along the value chain.

In the area of private households, the German Federal Government has commissioned the department in charge of the research of consumes to identify new representative figures for the collection of food waste. By means of household-books, referred to as the permissible survey method in the European Union (EU) methodological manual, the throw-away procedures are recorded in representatively selected households. The survey will be carried out over a year in several sessions, so that seasonal differences can be intercepted.

2.2 The Food Sharing Organisation in Germany: The Magdeburg Experience

With specific relation to Germany, food sharing experiences have been carried out and continually implemented since 2012 by means of the 'Foodsharing' organisation (https://foodsharing.de/). Its activity is declared as the contrast against food waste, which 'saves' possibly discarded food products (Sect. 1.5).

Nowadays, more than 200,000 registered users and over 25,000 volunteers—the so-called foodsavers—are giving to this association the face of an international movement, despite the organisation is based in a little multinational area involving Germany, Austria, and Switzerland. The number of farms that collaborate exceeds 3,000 units, with 22,764,237 kg of food already being rescued from waste, with clear growing tendency. Indeed, there are about 1,000 additional pickups taking place daily.[1] The voluntary commitment is the core of the foodsharing.de platform, meaning also that saving and sharing do not imply economic advantages. Therefore, food sharing is now considered as a no-profit and non-commercial association, without advertising.

The success of this platform can be also explained with the dimension of FW in Germany: each citizen is reported to waste more than 50 kg of food per year [13]. According to a World Wildlife Fund (WWF) environmental survey, Germans throw an average of 313 kg of edible food every second [14].

The goal of the organisation is to raise awareness around the issue as far as possible with the participation of all interested stakeholders. It is the only initiative of this size, currently based on volunteer engagement in this ambit, publicly communicating how many foodstuffs are thrown away, and suggesting different behaviours from a sustainability perspective. Moreover, the organisation supports legal solution based on French and Walloon models to reduce waste. The campaign is supported by foodsharing.de, Aktion Agrar, BUNDjugend, and Slow Food Youth.

Supermarkets and other food merchants might be obliged to convey all unsold/unsoldable and eatable foodstuffs commodities to dedicated associations such as foodsharing.de, or to their representatives, or customers. It has to be considered that food products declared unfit for human consumption would be bolstered to animals.

---

[1] The number of involved companies is 5392 units (date: 09 May 2019), according to the website: https://foodsharing.de/statistik.

In addition, other uses such as soil fertilisation or energy recovery (from burning or biogas ageing) should just be conceivable if the nourishment is not appropriate for people or animals. With relation to food commodities, interested products are picked up from common fridges which are placed in different strategic point, city by city.

An interesting experience is currently performed in Magdeburg by the EMMA Spielwagen e.V., a distribution point actively involved in the collection and distribution of food that would otherwise be wasted in garbage. The 'Emma' Children and Family Centre is a distribution point for food delivered by retailers or private individuals to prevent them from being destroyed. Boxes with bread, rolls, fruit, and vegetables are delivered during the week from volunteers. Anyone interested can take goods from publicly provided refrigerators, shelves, and boxes.

Since the second half of 2014, 10 tonnes of food have already been spared. 1800 people are active social members of a dedicated Facebook group; 50 volunteers help daily to collect and transport fruit, vegetables, bread, cakes, and other goods. Collected food products are transferred from a private consumer to another private citizen, on condition that each user unequivocally consents to inspect merchandise before consuming it, according to health safety criteria.

Interestingly, not only people with some economic disadvantage by use recovered foods picking up them in this Centrum. Many of interested 'stakeholders' use this system as aware strategy against food wastage.

In order to change this situation, Ralf Dounz-Weigt and a group of volunteers began to put refrigerators two years ago in the city of Magdeburg, from which everyone can pick up foods. These refrigerators are filled with unsold food from supermarkets and bakeries, but also with leftover from private kitchens. The entire system is firmly regulated. Involved volunteers sign up into an Internet datebook for trips. Moreover, possible help notices are posted in a Facebook group with 1900 individuals, with other advices concerning the availability of foods in coolers. Cleaning rules are likewise accessible, and the same thing is explicitly declared in eight languages when speaking of other rules, concerning minimum durability.

Consistently, people in charge to collect leftovers (food savers) visit shops, pastry kitchens, and markets in the city to get scraps: nourishment that has not been sold or will soon pass its expiration dates. They can pick whether (a) to keep leftovers themselves, (b) pass the nourishment along to others, or (c) take saved food products to one of the many '*Fairteiler*' (air-share) coolers and cabinets around the city, which are an active part of the movement.

Food sharing activities such as the above-described Magdeburg experience have likewise had accomplishment in different nations around the globe, including the USA and the UK. The model of utilising coolers has been replicated in Spain through 'solidarity ice fridges', bolstered by the leader of *Galdakao* in the northern Basque locale. With reference to the German sharing movement, Mr. Thurn—the author of the documentary 'Taste the Waste'—affirmed in 2014 that he collaborated with lawyers to ensure his food sharing plan didn't violate German or European regulations [15]. Some official worrying concerned the need of avoiding products with 'sell by' dates (meat and some dairy products), with a peculiar focus on those with 'best before' labels.

In Berlin, since the fridges are open to public access, authorities claim the food sharing fridges should be considerate as food businesses. According to this point of view, they must deal with EU Regulations—which are now being upheld by Berlin's Department for Consumer Protection. Indeed, the consumer protection authority recently issued hygiene regulations for food sausages. Now, for example, it is mandatory to keep a record of who is putting what into a refrigerator and getting it out. The reason is the necessity to identify some physical subject as responsible in the case some goods in the fridges would be rotten. Under this circumstance, both official authorities in charge of food safety and the person who took the food must have some interested person they can contact. In other cities such as Duisburg, food sharing subjects are in principle classified as private individuals; in this situation, authorities do not intervene. Retailers and producers have agreed on a legal base with the organisation that consumption of donated food is under the personal responsibility of the food sharing member [16]. This project has also received in 2015 two different awards in Germany with reference to environmental activities: the environmental prise of the state capital Magdeburg and a special award of the state of Saxony-Anhalt. In summary, it is a concrete and reliable history of success when speaking of active FW contrast in Germany and in the EU, also demonstrating that food sharing can be used as a fundamental part of FW-counterstrategies with other systems such as food recycling [17–35].

References

1. Parisi S, Barone C, Sharma RK (2016) Chemistry and food safety in the EU. Springer International Publishing, Cham. https://doi.org/10.1007/978-3-319-33393-9
2. Evans D (2012) Binning, gifting and recovery: the conduits of disposal in household food consumption. Environ Plan D Soc Space 30(6):1123–1137. https://doi.org/10.1068/d22210
3. Godemann J, Michelsen G (eds) (2011) Sustainability communication: interdisciplinary perspectives and theoretical foundations. Springer, Dordrecht
4. Heinrichs H (2013) Sharing economy: a potential new pathway to sustainability. Gaia 22(4):228–231
5. Davies AR, Doyle R (2015) Transforming household consumption: from backcasting to Home-Labs experiments. Ann Assoc Am Geogr 105(2):425–436. https://doi.org/10.1080/00045608.2014.1000948
6. Botsman R, Rogers R (2010) What's mine is yours: how collaborative consumption is changing the way we live, vol 5. Harper Collins, New York
7. Rifkin J (2000) The age of access. Penguin, Harmondsworth
8. Deutscher Bundestag (2017) Lebensmittelverschwendung verhindern, Drucksache 18/12631. http://dip21.bundes-tag.de/dip21/btd/18/126/1812631. Accessed 29 May 2019
9. BMUB/UBA (2017) Umweltbewusstsein in Deutschland 2016 - Ergebnisse einer repräsentativen Bevölkerungsumfrag. Bundesministerium für Umwelt, Naturschutz, Bau und Reaktorsicherheit (BMUB), Berlin, and Umweltbundesamt (UBA), Dessau-Roßlau
10. Eyerund T, Neligan A (2017) Verschwenderische Generationen X und Y, IW-Kurzbericht, No. 56.2017, Institut der deutschen Wirtschaft (IW), Köln. https://www.econstor.eu/handle/10419/167636. Accessed 29 May 2019
11. European Commission (2019) EU actions against food waste. European Commission, Brussels. https://ec.europa.eu/food/safety/food_waste/eu_actions_en. Accessed 27 May 2019

12. Anonymous (2019) Food waste: definition. EU Fusions, Brussels. https://www.eu-fusions.org/index.php/about-food-waste/280-food-waste-definition. Accessed 27 May 2019
13. Anonymous (2019) Germans waste 55 kg of food per person each year. The Local Europe AB, Stockholm. https://www.thelocal.de/20190220/german-government-announces-plans-to-curb-food-wastage. Accessed 28 May 2019
14. Anonymous (2019) WWF report: a third of German food lands in the trash. Deutsche Welle, www.dw.com. https://www.dw.com/en/wwf-report-a-third-of-german-food-lands-in-the-trash/a-18525699. Accessed 29 May 2019
15. Sarhaddi Nelson S (2014) Got leftovers to share? In Germany, there's a website for that. National Public Radio (NPR), Inc. https://www.npr.org/sections/thesalt/2014/06/27/321691095/got-leftovers-to-share-in-germany-theres-a-website-for-that?t=1
16. Lubeck J (2014) Viele Wege, Lebensmittel nicht zu verschwenden. Die Welt. http://www.welt.de/regionales/hamburg/article132545750/Viele-WegeLebensmittel-nicht-zu-verschwenden.html. Accessed 29 May 2019
17. Minsaas J, Heie AC (1979) Recycling of domestic food waste. Conserv Recycl 3(3–4):427–438. https://doi.org/10.1016/0361-3658(79)90040-74
18. Shaw P, Smith M, Williams I (2018) On the prevention of avoidable food waste from domestic households. Recycl 3(2):24. https://doi.org/10.3390/recycling3020024
19. Scholz K, Eriksson M, Strid I (2015) Carbon foot-print of supermarket food waste. Resour Conserv Recycl 94:56–65. https://doi.org/10.1016/j.resconrec.2014.11.016
20. Brancoli P, Rousta K, Bolton K (2017) Life cycle assessment of supermarket food waste. Resour Conserv Recycl 118:39–46. https://doi.org/10.1016/j.resconrec.2016.11.024
21. Refsgaard K, Magnussen K (2009) Household behaviour and attitudes with respect to recycling food waste—experiences from focus groups. J Environ Manag 90(2):760–771. https://doi.org/10.1016/j.jenvman.2008.01.018
22. Fehr M, Calcado MDR, Romao DC (2002) The basis of a policy for minimizing and recycling food waste. Environ Sci Policy 5(3):247–253. https://doi.org/10.1016/S1462-9011(02)00036-9
23. Da Cruz NF, Ferreira S, Cabral M, Simoes P, Marques RC (2014) Packaging waste recycling in Europe: is the industry paying for it? Waste Manag 34(2):298–308. https://doi.org/10.1016/j.wasman.2013.10.035
24. Comber R, Thieme A (2013) Designing beyond habit: opening space for improved recycling and food waste behaviors through processes of persuasion, social influence and aversive affect. Pers Ubiquit Comput 17(6):1197–1210. https://doi.org/10.1007/s00779-012-0587-1
25. Fait M, Scorrano P, Mastroleo G, Cillo V, Scuotto V (2019) A novel view on knowledge sharing in the agri-food sector. J Knowl Manag. https://doi.org/10.1108/jkm-09-2018-0572
26. Shirado H, Iosifidis G, Tassiulas L, Christakis NA (2019) Resource sharing in technologically defined social networks. Nat Commun 10(1):1079. https://doi.org/10.1038/s41467-019-08935-2
27. Sidali KL, Kastenholz E, Bianchi R (2015) Food tourism, niche markets and products in rural tourism: combining the intimacy model and the experience economy as a rural development strategy. J Sustain Tour 23(8–9):1179–1197. https://doi.org/10.1080/09669582.2013.836210
28. Marovelli B (2019) Cooking and eating together in London: food sharing initiatives as collective spaces of encounter. Geoforum 99:190–201. https://doi.org/10.1016/j.geoforum.2018.09.006
29. Hamari J, Sjöklint M, Ukkonen A (2016) The sharing economy: Why people participate in collaborative consumption. J Assoc Inf Sci Technol 67(9):2047–2059. https://doi.org/10.1002/asi.23552
30. Pascucci S, Dentoni D, Lombardi A, Cembalo L (2016) Sharing values or sharing costs? Understanding consumer participation in alternative food networks. NJAS Wagening J Life Sci 78:47–60. https://doi.org/10.1016/j.njas.2016.03.006
31. Tornaghi C (2017) Urban agriculture in the food-disabling city: (re)defining urban food justice, reimagining a politics of empowerment. Antipode 49(3):781–801. https://doi.org/10.1111/anti.12291

32. Falcone PM, Imbert E (2017) Bringing a sharing economy approach into the food sector: the potential of food sharing for reducing food waste. In: Morone P, Papendiek F, Tartiu V (eds) Food waste reduction and valorisation. Springer, Cham. https://doi.org/10.1007/978-3-319-50088-1_10
33. Kaushal LA (2018) The rise in the sharing economy: Indian perspective. In: Zakaria N, Kaushal LA (eds) Global entrepreneurship and new venture creation in the sharing economy. IGI Global, Hershey. https://doi.org/10.4018/978-1-5225-2835-7.ch007
34. Loh P, Agyeman J (2019) Urban food sharing and the emerging Boston food solidarity economy. Geoforum 99:213–222. https://doi.org/10.1016/j.geoforum.2018.08.017
35. Nishino N, Takenaka T, Takahashi H (2017) Manufacturer's strategy in a sharing economy. CIRP Ann 66(1):409–412. https://doi.org/10.1016/j.cirp.2017.04.004

Chapter 3
Food Sharing and Durable Foods. The Analysis of Main Chemical Parameters

Abstract The current market of foods and beverages worldwide is obliged to face two counteracting factors: the intensive production flow of many food and beverage commodities, year by year, and the concomitant loss of remarkable amounts of globally produced products. Food losses may be also confused with 'food waste', but these words usually concern two different phenomena. Food losses concern the final destination (food and beverage consumers), while food waste is related to the management of food supply chains. From a general viewpoint, food loss is sometimes unavoidable, while food waste should be adequately contrasted. Both phenomena seem to depend mainly on anthropic behaviours. However, the distinction between perishable and non-perishable foods should be considered as a critical factor because of Parisi's first law of food degradation. Apparently, food loss and waste cannot be predicted and adequately contrasted without a strong analysis concerning the nature of involved foods, even if the active subject—the food consumer—is the only aware player in this ambit. This chapter examines food products that can be easily collected because of their supposed long shelf life, with relation to the most observed and analysed chemical–physical features (for food safety evaluation).

Keywords Durability · European Union · Food loss · Food sharing · Food waste · Moisture · Water

Abbreviations

EU	European Union
FAO	Food and Agriculture Organization of the United Nations
F&B	Food and beverage
FCRN	Food and Climate Research Network
FL	Food loss
FSC	Food supply chain
FW	Food waste
FFV	Fresh fruits and vegetable
OECD	Organization for Economic Co-operation and Development

© The Author(s), under exclusive license to Springer Nature Switzerland AG 2019
A. Pellerito et al., *Food Sharing*, Chemistry of Foods,
https://doi.org/10.1007/978-3-030-27664-5_3

USA   United States of America
UK   United Kingdom
WRAP Waste and Resources Action Programme

3.1   Does Food Sharing Concern All Foods and Beverages?

At present, the current market of foods and beverages worldwide is obliged to face two counteracting factors:

(a)   The intensive production flow of many food and beverage (F&B) commodities, year by year, with the resulting flow of F&B items along different logistic and consumer chains;
(b)   The concomitant loss of remarkable amounts of the globally produced F&B items.

Food losses may be also confused with 'food waste', but these words usually concern two different phenomena. According to the Food and Agriculture Organization of the United Nations (FAO) definition, 'food loss' (FL) represents 'any change in the availability, edibility, wholesomeness or quality of edible material that prevents it from being consumed by people' [1]. On the other side, the European Project 'FUSIONS' defined food waste (FW) as 'any food, and inedible parts of food, removed from the food supply chain to be recovered or disposed (including composted, crops ploughed in/not harvested, anaerobic digestion, bio-energy production, co-generation, incineration, disposal to sewer, landfill or discarded to sea)' [2]. Consequently, FL has to be considered with relation to the final destination (food and beverage consumers), while FW concerns the food supply chain (FSC) and related food business operators (FBO) and other interested stakeholders. From a general viewpoint, FL is sometimes unavoidable, while FW should be adequately contrasted in all possible situations, with reference also to associated costs [2–5]. As discussed in Chap. 1, apparently, countries with medium and high-income values show a general FW level of 95–115 kg per person each year, while developing countries appear to waste yearly just 6–11 kg per person year [6].

Consequently, geographical locations and geopolitics in general may influence food waste. The main FW amount in developing countries (>40%) is reported to occur in the first production stages, after harvest and during processing steps. On the other hand, FW seems more abundant at the retail and consumer level when speaking of industrialised countries (FW $\geq$ 40%) [7].

Post-harvest losses are in part due to the available technology in a country, and additionally to the degree to which agricultural products are requested in the market. The three main driving key factors in the FSC organisation and future trends in developing countries seem to be:

(1) Urbanisation and contraction of the agricultural sector [8].
(2) Dietary transition and increment in family unit earnings, especially in Brazil, Russia, India and China (BRIC) Nations. In particular, these social phenomena appear to confirm Bennett's Law [9], where an increment in income rate leads to a decrement in food share of starchy staples. The preference of vulnerable food, with shorter shelf life, ends up with a more prominent FW production and a greater impact on the environment [10].
(3) Increased trade globalisation [11].

The most part of people living in rural areas depends on short FSC, relying on few post-harvest infrastructure and technologies. Extended FSC networks can provide food to people living in urban areas, including generally many intermediate steps between cultivators and purchasers. On the other side, farming mostly takes place on a small scale with different degrees of engagement in local markets, and a fast decreasing number of subsistence farmers who neither purchase nor sell basic food [12]. As a result, food costs tend to rise if production is excessive, and some counterstrategy is needed and sometimes implemented [13]. Another distinct factor is represented from retail agents: stores are the prevailing mediator between ranchers and customers. Indeed, even in poorer transitional economies, grocery stores are the fundamental vehicle for conveying enhanced eating regimens. This is altogether reliant on foreign direct investment, with high development rates in Eastern Europe, Asia and Latin America [14].

Other difficulties connected with FL and FW phenomena in developing countries can be listed as follows:

(a) Payment terms represent a source of discouragement for small growers;
(b) Quality standards applied by retailers on products defect smallholders from providing products to the market;
(c) Heavy penalties, according to signed contracts, for partially or totally undelivered orders by providers have to be considered;
(d) The restitution of items, according to signed contracts (retailers can give back F&B articles to providers once a leftover timeframe of realistic usability has been reached) has to be taken into account;
(e) Frequently inadequate demand predictions and replacement systems should be evaluated, with some questionable points when speaking of FSC transparency.

On the other hand, FL and FW phenomena are differently observed in developed countries (including European member states), but a common point is always found: the exit of F&B items happens mainly towards the end of the FSC. The interested steps are defined 'appropriation', 'retail', and 'last utilisation' by the final consumer [15]. In countries with a medium–high-income per capita, consumers seem often suppose that improved resource efficiency with reduced FW production can be obtained just through processing food in a central system. Even if the majority of FW is produced in factories, it should be supposed that a reliable FW reduction is overall achieved with 'polite' home behaviours.

Anyway, FW and FL seem to depend mainly on anthropic behaviours. However, the passive subject in the FSC—F&B items—has not yet analysed in detail. In the context of FSC losses, it is important to keep in consideration of the distinction between perishable and non-perishable foods and the adequacy of FSC networks. This evaluation is critical when speaking of FL and/or FW predictions because of the Parisi's first law of food degradation which clearly states that all food products are inevitably subjected over time to a continuous and unstoppable transformation of their chemical, physical, microbiological, sensorial and structural features [16–19].

A simple example concerning the sector of cereals may help. Grain losses happen in post-harvest systems as a result of physical losses (spillage, consumption by pests) or quality reduction. Most of the performed FW/FL studies are based on practical parameters because the quantitative measurement of quality reductions in grains may be estimated with some difficulty. However, the main part of the worldwide production of cereals needs to be kept in storage (1 month to more than 12 months); this fact can be used to monitor indirectly FL concerning stored grains.

The main factors leading to overestimation of grain losses may be:

(a)   The evaluation of extreme values instead of the average estimation;
(b)   The elimination of partially-damaged grains, when this raw material would rather be utilised by farmers for their own utilization or as animal feed.

In addition, grain losses are counted twice in different steps in the post-harvest system.

With reference to post-harvest losses (for perishable crops), the example of cereal grains is not suitable because of the demonstrated tendency of perishable crops to spoil. Horticultural products usually tend to spoil in industrialised and developing countries, even if at different stages in the FSC. Basic reasons for such an evident spoilage may be different. Kader assessed that around 33% of produced fresh fruits and vegetables (FFV) worldwide are lost before it can be consumed by customers. Losses in the United States of America (USA) are evaluated from 2 to 23%, depending on the product, with an overall average of 12% [20]. FFV losses in the United Kingdom (UK) are estimated to be around 9%, but this estimation does not take into account produce that might be left on fields because they do not satisfy cosmetic or quality criteria [21].

In the European Union (EU), the recent regulatory situation has removed certain quality and size criteria for trade FFV with the resulting suitability for many of these commodities when speaking of the EU market [22]. Anyway, and similarly for grains, there is confirmation of overestimation of perishable product losses. There are generally few, 'just-in-time' managed and distant steps from the field to the buyer. Conventional collecting strategies, e.g. utilising sticks to collect papaya and mango, may damage organic products, leading to losses for more broadened supply chains. Anyway, a high FFV portion is utilised by local consumers.

This discussion seems to demonstrate that FW and FL cannot be predicted and adequately contrasted without a strong analysis concerning the nature of involved foods, even if the active subject—the food consumer—is the only aware player in this ambit.

3.2 Can Durability Be the Key with Reference to Food Sharing?

With reference to FW, existing and available surveys have revealed that a significant amount comes from perishable products such as meat and poultry, FFV, and beverage items. Interestingly, related FW amount is mainly represented by premade food products that remain unsold and by-products: dairy, eggs, meat, fish, meals, baker, drinks, and FFV above all [23–25].

Obviously, the anthropic factor—in terms of human behaviour—is the first of examined points when speaking of FW worldwide. However, the progressive tendency of food consumers towards the purchasing activity for many types of F&B items with a negligible amount (in terms of weight or volumetric capacity) should be considered. As a result, the higher the number of different items on the market and—broadly—in the complete FSC, the higher the danger of wasted products because of the loss of acceptability requirements during shelf life. It should be noted that the last words—shelf life—may be one of the solutions when speaking of FW countermeasures.

In other words, should shelf life be extended for a generic food or beverage, the FW amount referred to a huge number of F&B products of the same type would easily decrease because of the theoretical resistance of F&B items over time. Consequently, one of the maximum FW key-driver factors—expiration—would diminish, while other factors would naturally become important (Chap. 1). On the other side, there is a regulatory consideration: foods can be placed on the market provided that safety is assured, including the needed acceptability of F&B products on various levels (microbiological profiles, sensorial acceptability, usability, etc.). For these reasons, shelf life should be always evaluated and demonstrated [19].

In summary:

(a) Shelf-life extension can measurably reduce FW impact, and
(b) Perishable F&B products are probably exposed to the risk of improper purchasing (Chap. 1) with consequent increase of probable wasted units.

This discussion does not consider FL in each step of production processes, highlighting the importance of consumers' behaviours and different retail strategies (Chap. 1). As a consequence, and with particular relation to food sharing activities, shareable products would easily be collected and distributed on the basis of demonstrated high durability. This estimation highlights also the role of main chemical parameters, food by food, which should be considered in advance.

3.3 Food Typologies and Main Chemical Parameters

Figure 3.1 shows the most important categories of F&B products which may be considered as 'high FW-risk' products:

Fig. 3.1 Most important
categories of F&B products
which may be considered as
'high FW-risk' products, at
present

- Dairy foods;
- Eggs;
- Meat and meat preparations;
- Fish, seafood, and related preparations;
- Prepared meals;
- Bakery products;
- Drinks;
- FFV.

As discussed in Sect. 3.1, the main reasons for FW by food consumers could be interpreted as the direct or indirect consequence of their high-perishability or perishability status, provided that an important prerequisite—adequate storage temperatures and conditions—is considered a priori. As discussed in Sect. 3.2, food sharing systems are fridges and coolers, although it has been agreed in Germany that consumption of donated foods is under the personal responsibility of the food sharing member [26].

This discussion does not consider FL in each step of production processes, high-lighting the importance of consumers' behaviours and different retail strategies (Chap. 1). As a consequence, and with particular relation to food sharing activities, shareable products would easily be collected and distributed on the basis of demonstrated high durability. This estimation highlights also the role of main chemical parameters, food by food, which should be considered in advance.

However, F&B categories shown in Fig. 3.1 should always be considered. What about the main chemical parameters which should be evaluated because of their influence on durability (or possible monitoring actions)? The following section discusses briefly the main chemical parameters which should be considered by a food sharing perspective (because excessive values for these products may recommend their exclusion from these social actions). Some peculiar product with claimed long durabilities may be excluded from this discussion, such as certain vegetable oils.

3.3.1 Water (or Moisture) and Water Activity

Water means 'life' when speaking of living microorganisms and superior life forms. It is an excellent solvent, able to react itself in many chemical and biochemical reactions, and it is ranked first in many food and beverage products, although its abundance is not always so prominent [27]. For this main reason, the reduction of water is generally linked with associated long shelf-life performances in many foods.

Because of the importance of such a bioavailable solvent for organic life forms, the amount of bioavailable water is critical. On the other hand, increased water amounts mean also that all dissolved matters (proteins, carbohydrates, mineral substances, gases, etc.) in foods and beverages are readily bioavailable for spoilage and pathogen microorganisms, while the hydrophobic matter—lipids—might be considered as a chemical safeguard against microbial degradation. An important exception concerns the oxidation and similar chemical reactions on fatty chains. However, the durability of foods may be correlated with the quantity of the water-soluble fraction of a whole food. Consequently, water is the key factor when speaking of perishable and highly perishable foods; on the contrary, low water amounts mean probably extended durability.

According to many scientific papers, water—analysed under the name 'moisture'[1]—should be removed at the most possible level. Table 3.1 shows abovementioned food and beverage categories with normal water amounts and the suggested quantity of residual moisture which could justify long-durability performances. These data are shown here as simple examples, meaning that extended durability often corresponds to canned, or dried, or frozen products.

On the other hand, the best correlation between water presence and durability would be obtained when speaking of water activity [27]. This measure would be more useful than water or moisture determination: water activity values between 0.6 and 0.9 mean that bioavailable water for microbial spoilage and chemical reactions is surely assured if compared with water activity values of 0.2–0.4, when only a few life forms such as osmophilic microorganisms can survive. However, the problem is that food products are often evaluated by means of simple chemical parameters such as moisture. For this reason, water activity is not always considered as the first choice when speaking of durable foods (or perishable products). On the other side, heating and concentration processes are often evaluated 'just-in-time' with simple moisture determinations instead of water activity estimations [19].

Table 3.1 is only a small example of different water amounts in normal (perishable) foods and in durable products, showing visually how food sharing managers should evaluate the choice of more durable and convenient foods, provided that adequate storage conditions are assured.

Certain foods may have many different versions, depending on chemical features and processing options. Cheeses may be interesting enough, in this ambit. Chap. 4 is

---

[1] Moisture does not mean 'water'. On the contrary, it may be defined as the amount of matter which could be removed under heating, into a dedicated oven, or by means of irradiation systems such as infra-red balances or microwave ovens.

Table 3.1 Water in selected foods and beverages: normal values and reduced moisture amounts when speaking of long-durability foods (fit for food sharing purposes) [28, 29]

Normal foods	Moisture content (%)	Long-durability foods	Moisture content (%)
Dairy foods			
Cow milk	87	Milk powder	5
Unsalted butter	16–18	Butter oil, anhydrous	0.24
Cheese (various types)	30–65	Dried cheese (various types)	<30–35
Eggs	75	Egg, whole, dried	2.8
Meat and meat preparations	65–75	Meat and meat preparations	
Beef, cured, luncheon meat, jellied	74.60	Frankfurter	56.31
Fish, seafood, and related preparations			
Fish, anchovy, European, raw	73.37	Fish, anchovy, European, canned in oil, drained solids	50.30
Prepared meals			
Soup, stock, beef, home-prepared	95.89	Soup, bean with frankfurters, canned, condensed	67.70
Bakery products			
Bread	35–38	Bread, sticks, plain	6.10
Wheat flour, white, bread, enriched	13.36	Wheat flour, white, all-purpose, enriched, bleached	12–14
Drinks			
Beverages, grape drink, canned	84.20	Beverages, fruit punch juice drink, frozen concentrate	55.50
FFV (average value)	70–90	FFV (average value)	
Apricots, raw	86.4	Apricots, dried, sulphured, uncooked	30.9

dedicated to the evaluation of certain cheeses by a food sharing and food-recycling perspective, taking into consideration two important aspects of the problem: (a) economic value and (b) durability options. Food sharing and reuse can co-exist and be valuable options against food waste [28–46].

References

1. FAO (1981) Food loss prevention in perishable crops. FAO Agricultural Services Bulletin 43. The Food and Agriculture Organisation of the United Nations, Rome
2. Anonymous (2019) Food waste: definition. EU Fusions, Brussels. Available https://www.eu-fusions.org/index.php/about-food-waste/280-food-waste-definition. Accessed 27 May 2019
3. Slow Food (2019) Food waste. Slow Food, Bra. Available https://www.slowfood.com/sloweurope/en/topics/food-waste/. Accessed 27 May 2019
4. FAO (2017) EU and FAO bring combined weight to bear on food waste, antimicrobial resistance. Food and Agriculture Organization of the United Nations (FAO), Rome. Available http://www.fao.org/news/story/it/item/1040628/icode/. Accessed 27 May 2019
5. Stenmarck Å, Jensen C, Quested T, Moates G (2016) Estimates of European food waste levels. Available http://www.eu-fusions.org/phocadownload/Publications/Estimates%20of%20European%20food%20waste%20levels.pdf. Accessed 27 May 2019
6. Godemann J, Michelsen G (eds) (2011) Sustainability communication: interdisciplinary perspectives and theoretical foundations. Springer, Dordrecht
7. Gustavsson J, Cederberg C, Sonesson U, van Otterdijk R, Meybeck A (2011) Global food losses and food waste—extent, causes and prevention. Food and Agriculture Organization of the United Nations (FAO), Rome. Available http://www.fao.org/3/a-i2697e.pdf
8. United Nations (2015) World urbanization prospects—the 2014 revision, ST/ESA/SER.A/366. United Nations, Department of Economic and Social Affairs, Population Division, New York. Available https://population.un.org/wup/Publications/. Accessed 30 May 2019
9. Bennett MK (1941) International contrasts in food consumption. Geogr Rev 31:365–376
10. Lundqvist J, de Fraiture C, Molden D (2008) Saving water: from field to fork: curbing losses and wastage in the food chain. Stockholm International Water Institute, Stockholm
11. Henderson DR, Handy C, Neff SA (1997) Globalization of the processed foods market, agricultural economic report No. 742. Food and Consumer Economics Division, Economic Research Service, U.S, Department of Agriculture, Washington, DC
12. Jayne TS, Zulu B, Nijhoff JJ (2006) Stabilizing food markets in eastern and southern Africa. Food Policy 31(4):328–341. https://doi.org/10.1016/j.foodpol.2006.03.008
13. Hattersley L, Isaacs B, Burch D (2013) Supermarket power, own-labels, and manufacturer counterstrategies: international relations of cooperation and competition in the fruit canning industry. Agric Hum Values 30(2):225–233. https://doi.org/10.1007/s10460-012-9407-5
14. Reardon T, Timmer C, Berdegué JAP (2007) Super-market expansion in Latin America and Asia, implications for food marketing systems new directions in global food markets. United States Department of Agriculture (USDA), Economic Research Service, Washington
15. Griffin M, Sobal J, Lyson TA (2009) An analysis of a community food waste stream. Agric Hum Values 26(1):67–81
16. Steinka I, Barone C, Parisi S, Micali M (2017) Technology and chemical features of frozen vegetables. The chemistry of frozen vegetables. Springer, Cham, pp 23–29
17. Pellerito A, Ameen SM, Micali M, Caruso G (2018) Antimicrobial substances for food packaging products: the current situation. J AOAC Int 101(4):942–947. https://doi.org/10.5740/jaoacint.17-0448
18. Volpe MG, Stasio MD, Paolucci M, Moccia S (2015) Polymers for food shelf-life extension. Functional polymers. In: Cirillo G, Spizzirri UG, Iemma F (eds) Food science—from technology to biology, vol 1. Wiley, Hoboken
19. Parisi S (2002) Profili evolutivi dei contenuti batterici e chimico-fisici in prodotti lattiero-caseari. Ind Aliment 41(412):295–306
20. Kader AA (2005) Increasing food availability by reducing postharvest losses of fresh produce. Acta Hortic 682:2169–2175. https://doi.org/10.17660/ActaHortic.2005.682.296
21. Garnett T (2006) Fruit and vegetables and UK greenhouse gas emissions: exploring the relationship, FCRN working paper 06-01 Rev. A. Food and Climate Research Network (FCRN), University of Surrey, Guildford, 134pp

22. OECD (2012) Peer review of the fruit and vegetables quality inspection system in the Netherlands. Organization for Economic Co-Operation and Development (OECD) Scheme for the Application of International Standards for Fruit and Vegetables, p 15
23. Lee P, Peter W, Hollins O (2010) Waste arisings in the supply of food and drink to households in the UK. Waste and Resources Action Programme (WRAP), Banbury
24. Pekcan G, Köksal E, Küçükerdönmez Ö, Özel H (2006) Household food wastage in Turkey. Food and Agriculture Organization of the United Nations (FAO), Rome
25. Quested T, Johnson H (2009) Household food and drink waste in the UK. Waste and Resources Action Programme (WRAP), Banbury
26. Lubeck J (2014) Viele Wege, Lebensmittel nicht zu verschwenden. Die Welt. Available http://www.welt.de/regionales/hamburg/article132545750/Viele-WegeLebensmittel-nicht-zu-verschwenden.html. Accessed 29 May 2019
27. Belitz HD, Grosch W, Schieberle P (2009) Food chemistry, 4th edn. Springer, Berlin
28. Minsaas J, Heie AC (1979) Recycling of domestic food waste. Conserv Recycl 3(3–4):427–438. https://doi.org/10.1016/0361-3658(79)90040-74
29. Shaw P, Smith M, Williams I (2018) On the prevention of avoidable food waste from domestic households. Recycling 3(2):24. https://doi.org/10.3390/recycling3020024
30. Scholz K, Eriksson M, Strid I (2015) Carbon foot-print of supermarket food waste. Resour Conserv Recycl 94:56–65. https://doi.org/10.1016/j.resconrec.2014.11.016
31. Brancoli P, Rousta K, Bolton K (2017) Life cycle assessment of supermarket food waste. Resour Conserv Recycl 118:39–46. https://doi.org/10.1016/j.resconrec.2016.11.024
32. Refsgaard K, Magnussen K (2009) Household behaviour and attitudes with respect to recycling food waste—experiences from focus groups. J Environ Manag 90(2):760–771. https://doi.org/10.1016/j.jenvman.2008.01.018
33. Fehr M, Calcado MDR, Romao DC (2002) The basis of a policy for minimizing and recycling food waste. Environ Sci Policy 5(3):247–253. https://doi.org/10.1016/S1462-9011(02)00036-9
34. Da Cruz NF, Ferreira S, Cabral M, Simoes P, Marques RC (2014) Packaging waste recycling in Europe: is the industry paying for it? Waste Manag 34(2):298–308. https://doi.org/10.1016/j.wasman.2013.10.035
35. Comber R, Thieme A (2013) Designing beyond habit: opening space for improved recycling and food waste behaviors through processes of persuasion, social influence and aversive affect. Pers Ubiquit Comput 17(6):1197–1210. https://doi.org/10.1007/s00779-012-0587-1
36. Fait M, Scorrano P, Mastroleo G, Cillo V, Scuotto V (2019) A novel view on knowledge sharing in the agri-food sector. J Knowl Manag. https://doi.org/10.1108/jkm-09-2018-0572
37. Shirado H, Iosifidis G, Tassiulas L, Christakis NA (2019) Resource sharing in technologically defined social networks. Nat Commun 10(1):1079. https://doi.org/10.1038/s41467-019-08935-2
38. Sidali KL, Kastenholz E, Bianchi R (2015) Food tourism, niche markets and products in rural tourism: combining the intimacy model and the experience economy as a rural development strategy. J Sustain Tour 23(8–9):1179–1197. https://doi.org/10.1080/09669582.2013.836210
39. Marovelli B (2019) Cooking and eating together in London: food sharing initiatives as collective spaces of encounter. Geoforum 99:190–201. https://doi.org/10.1016/j.geoforum.2018.09.006
40. Hamari J, Sjöklint M, Ukkonen A (2016) The sharing economy: why people participate in collaborative consumption. J Assoc Inf Sci Technol 67(9):2047–2059. https://doi.org/10.1002/asi.23552
41. Pascucci S, Dentoni D, Lombardi A, Cembalo L (2016) Sharing values or sharing costs? Understanding consumer participation in alternative food networks. NJAS Wageningen J Life Sci 78:47–60. https://doi.org/10.1016/j.njas.2016.03.006
42. Tornaghi C (2017) Urban agriculture in the food-disabling city: (re)defining urban food justice. Reimagin Polit Empower Antipod 49(3):781–801. https://doi.org/10.1111/anti.12291
43. Falcone PM, Imbert E (2017) Bringing a sharing economy approach into the food sector: the potential of food sharing for reducing food waste. In: Morone P, Papendiek F, Tartiu V (eds) Food waste reduction and valorisation. Springer, Cham. https://doi.org/10.1007/978-3-319-50088-1_10

44. Kaushal LA (2018) The Rise in the sharing economy: Indian perspective. In: Zakaria N, Kaushal LA (eds) Global entrepreneurship and new venture creation in the sharing economy. IGI Global, Hershey. https://doi.org/10.4018/978-1-5225-2835-7.ch007
45. Loh P, Agyeman J (2019) Urban food sharing and the emerging Boston food solidarity economy. Geoforum 99:213–222. https://doi.org/10.1016/j.geoforum.2018.08.017
46. Nishino N, Takenaka T, Takahashi H (2017) Manufacturer's strategy in a sharing economy. CIRP Ann 66(1):409–412. https://doi.org/10.1016/j.cirp.2017.04.004

Chapter 4
Food Waste and Correlated Impact in the Food Industry. A Simulative Approach

Abstract The phenomenon of food waste appears generated and highly relevant in many industrialised countries with reference to perishable products such as meat and poultry, fruits and fresh vegetables, beverage items, and milk/dairy foods. Other perishable or highly perishable food products—eggs, fish, prepared meals, bakery products—should be considered in this ambit. The higher the number of different items on the market and in the complete food supply chain, the higher the danger of wasted products because of the loss of acceptability requirements during shelf life. Consequently, adequate countermeasures against food waste should take into account the problem of perishability. On the other side, and by a regulatory viewpoint, foods on the market have to be safe, legal, and correspondent to the claim(s) reported on labels. From a practical and industrial perspective, food waste could suggest the reuse and recycling of foods and beverages, provided that certain and minimal safety requirements are maintained. This chapter shows a simulation concerning a particular food product—a melted cheese—in five versions. Recycling may be good enough when speaking of food waste countermeasures and price reduction, but there is no demonstration that needful people can effectively take advantage of this system because food recycling occurs usually in a few and limited areas. On the contrary, On the other hand, food sharing should be considered as a distinctive advantage for food consumers as part of a more complex and multi-operational strategy.

Keywords Durability · Food sharing · Residual shelf life · Moisture · Processed cheese · Recycling · Social supermarket

Abbreviations

FAO	Food and Agriculture Organization of the United Nations
FSC	Food supply chain
FW	Food waste
FFV	Fresh fruits and vegetable
MFFB	Moisture on free-fat basis
RSL	Residual shelf life

SSM   Social supermarket
WRAP   Waste and Resources Action Programme

4.1 Foods and Beverages Containing Recycled Foods

The phenomenon of food waste (FW) appears generated and highly relevant in many industrialised countries (Chap. 3) with relation to perishable products such as meat and poultry, fruits and fresh vegetables (FFV), beverage items, and milk/dairy foods. Other perishable or highly perishable food products—eggs, fish, prepared meals, bakery products—should be considered in this ambit [1–3].

It has also been noted that the higher the number of different items on the market and in the complete food supply chain (FSC), the higher the danger of wasted products because of the loss of acceptability requirements during shelf life. Consequently, adequate FW countermeasures should take into account the problem of perishability. On the other side, and by a regulatory viewpoint, foods on the market have to be safe, legal, and correspondent to the claim(s) reported on labels.

What about the impact of expired foods and beverages in the FSC? Naturally, the shorter the shelf-life performance for a particular product (or product category), the higher the predictable risk and the amount of wasted foods. From a practical and industrial perspective, FW could suggest the reuse and recycling of foods and beverages, provided that certain and minimal safety requirements are maintained. In general, this situation means that foods and beverages with the 'sell-by-date' mention may be evaluated for further processing and/or reuse and recycling in the food and beverage industry some day before the mentioned date. On the other hand, legally expired foods should not be considered.

This chapter shows a simulation concerning a particular food product—a melted cheese—in five versions. The first version is a melted cheese with normal raw materials and additives without possible reuse or recycling of alternative cheeses or ingredients originally placed on the market. On the other side, versions 2–5 are melted cheeses containing one specified melted cheese originally on the market and with three days left after its 'sell-by-date' mention (different amounts).

4.2 Foods and Beverages Containing Recycled Foods. Economic Savings

Melted cheeses can have different formulations. A dedicated simulation can rely on the following list of ingredients:

- Water;
- Cheeses;
- Vegetable oil and/or animal butter;
- Rennet casein;
- Food-grade salt;
- Melting agent(s) such as sodium citrate;
- Acidity corrector(s) such as citric acid.

On these bases, the following cheese formulation may be realised:

- Water: 50%;
- Cheeses: 20%;
- Vegetable oil and/or animal butter: 17%;
- Rennet casein: 10%;
- Food-grade salt: 1.3%;
- Melting agent(s) such as sodium citrate: 1.2%;
- Acidity corrector(s) such as citric acid: 0.5%.

This list is completely hypothetical, serving only for simulation purposes. Should a food technologist evaluate the above-shown formulation, he would also consider that the 'cheeses' ingredient is a relevant component (20%). In a normal situation, cheeses produced into a cheese-making factory may be reused within some day from the original production date without excessive problems. However, economic considerations should be taken into account because each recycled raw material into a factory has a certain economic impact (as unsold good). In fact, each food preparation has a related cost determined as the sum of used raw materials and additives and other costs related to its production. Consequently, food technologists would find it more convenient to search for certain raw materials—such as cheeses placed on the market with a few days of shelf-life expectation—instead of 'reworking' or 'offline' materials produced into the same factory [4, 5].

Moreover, cheeses produced after some days tend to have bad or medium performances when speaking of melting attitude and water absorption. In general, 'young' cheeses have bad melting attitudes. On the contrary, 'old' cheeses tend to melt very well and with acceptable water absorption, on condition that a certain amount of rennet casein is provided. However, it should be considered that the aqueous amount into cheeses for reworking operations may be partially lost (expelled from the intermediate mass which will become the final product) because the protein fraction is partially demolished and consequently unable to absorb all the initial water quantity.

As a result, the simulation may have two distinct versions as shown in Table 4.1. From the economical viewpoint, it may be assumed that the final product—Cheese No. 1—means a final production price of 4.00 EUR, and the same thing is affirmed for the ingredient 'cheeses' on condition that this ingredient is identical to the same product. On the other hand, the alternative version—Cheese No. 2—would use recycled cheeses (from the market) with a final price (production, transportation, accessory costs, etc.) of 2.80 EUR. Consequently, 0.24 EUR would be saved when speaking of Cheese No. 2 with a final production price of 3.76 EUR (−6%).

Table 4.1 A simulation concerning melted cheeses

Ingredients	Cheese No. 1	Cheese No. 2
	Amount (%)	Amount (%)
Water	50	50
Cheeses	20	00
Reused cheeses (from the market, three days left)	0	20
Vegetable oil	17	17
Rennet casein	10	10
Food-grade salt	1.3	1.3
Melting agent(s) such as sodium citrate	1.2	1.2
Acidity corrector(s) such as citric acid	0.5	0.5
Economic value	4.00 EUR	3.76 EUR

Two possible versions, where the difference is the use of 'old' cheeses with three days left (with reference to the original 'sell-by-date' mention)

With relation to safety and legality, there are not particular observations except for the needed evaluation of recycled cheeses before production. By the technological viewpoint, the lower amount of adsorbed water in Cheese No. 2 may be assumed negligible, while meltability is generally improved.

This very simple simulation can demonstrate that food recycling of edible foods and beverages on the market is a valuable resource for industries. Because of the need to reduce FW and food losses along the FSC, the same simulation may also demonstrate that reusable foods and beverages can be also good and acceptable for food sharing activities because of their 'saving power' when speaking of economic values. The observed saving performance in our simulation (–6%) could be correlated with the possible amount of FW reduction in food markets. Naturally, more research is needed in this complicated ambit before confirming similar correlations.

However, the ambit of FW countermeasures should require other analyses because of the important point of durability options (Sect. 3.3.1). As displayed in Table 3.1, there are 'normal' and 'long-durability' foods which could be used when speaking of food sharing activities, but the general trend is in favour of long-durability products. In particular, 'food banks'—no-profit associations that gather sustenance from retailers and redistribute it to poor people for free—redistribute edible goods to or provide suppers for supporting poor people. In this ambit, edible commodities still fit for human consumption—although very close to the termination date or with some nonconformity to 'esthetical' or 'hedonistic' norms or performances—could be gathered and given to needful individuals requiring favoured access (Sect. 1.4.1). However, the basic problem is: Could all foods and beverages be redistributed? How about products with very limited residual shelf life (RSL)?

In addition, social supermarkets (Sect. 1.4.2) put unused and 'saved' food and beverage commodities for sale at emblematic costs to a limited community of individuals in danger of neediness. In general, only a restricted variety of nourishment and household products can be distributed, and the activity is directed only to documented individuals in danger of neediness. In addition—and this point is critical from the legal angle—shelf prices are brought down from 30 up to 70% if compared to conventional stores for two reasons, including also durability. In fact, natural waste (expired food) is not accepted and should be reused for other purposes. On the other side, donations should concern just items which have not reached their last utilisation date and can really be utilised (i.e. given over one day before the lapse date).

The basic and legal problem concerns the clarification of the 'last before date' meaning (Sect. 1.4.2). This date could be misused by retailers in an illegal way with their providers, for example, to refuse items if they have in over number. For this reason, legal countermeasures are needed. As a result, only legally and unexpired products could be managed by social supermarkets and destined for human consumption. Another (important) strategy concerns the reuse of food by-products not suitable for human utilisation which can be recycled to feed animals. On these bases, by-products destined for animal feeding can be separated from 'biowaste'.

The point of durability is justified because many food sharing options and activities may concern the use of refrigerators which can be open to public access, meaning also that the same activities might be confused with normal food businesses. A legal countermeasure may be the classification of food sharing subjects as private individuals; in this situation, official authorities do not intervene. In other words, retailers and producers can agree on a legal base with the interested food sharing organisation that consumption of donated food is under the personal responsibility of the food sharing member (Sect. 2.2).

Anyway, a common point concerning all food sharing activities—the use of food and beverage products with a restricted durability—has to be investigated. On the other hand, the economic importance of usable and recyclable foods and beverages should be considered as the opposite face of the food sharing activities. The following sections are dedicated to a simulative study of economic values and durability options with relation to processed cheeses, with some interesting conclusions.

4.3 Foods and Beverages Containing Recycled Foods. Durability Options

As discussed in Sect. 4.2, melted cheeses can have different formulations. The previous simulation—explained in Table 4.1—can basically rely on the following list of ingredients:

- Water;
- Cheeses;
- Vegetable oil and/or animal butter;
- Rennet casein;

- Food-grade salt;
- Melting agent(s) such as sodium citrate;
- Acidity corrector(s) such as citric acid.

The initial formulation for a processed cheese—as displayed in Table 3.1— assigns the first four positions to water (50% on the total amount of ingredients), cheeses (20%), vegetable oil and/or animal butter (17%), and rennet casein (10%). The remaining components account only for 3% of the total quantity of ingredients.

This list is completely hypothetical, serving only for simulation purposes. Should a food technologist evaluate the above-shown formulation, he would also consider that the 'cheeses' ingredient is a relevant component (20%). In a normal situation, cheeses produced into a cheese-making factory may be reused within some day from the original production date without excessive problems. However, economic considerations should be taken into account because each recycled raw material into a factory has a certain economic impact (as unsold good). In fact, each food preparation has a related cost determined as the sum of used raw materials and additives and other costs related to its production. Consequently, food technologists would find it more convenient to search for certain raw materials—such as cheeses placed on the market with a few days of shelf-life expectation—instead of 'reworking' or 'offline' materials produced into the same factory [4].

Moreover, cheeses produced after some days tend to have bad or medium performances when speaking of melting attitude and water absorption. In general, 'young' cheeses have bad melting attitudes. On the contrary, 'old' cheeses tend to melt very well and with acceptable water absorption, on condition that a certain amount of rennet casein is provided. However, it should be considered that the aqueous amount into cheeses for reworking operations may be partially lost (expelled from the intermediate mass which will become the final product) because the protein fraction is partially demolished and consequently unable to absorb all the initial water quantity.

Two distinct versions have been proposed in Table 4.1, with an economic result in favour of the second product—Cheese No. 2—because of a reduced price (3.76 EUR) if compared with Cheese No. 1 (4.00 EUR). Actually, the difference has been justified only with the use of recycled cheeses from the market (with a related price concerning also production, transportation, accessory costs, etc., of 2.80 EUR). Consequently, 0.24 EUR would be saved when speaking of Cheese No. 2.

However, some points have to be discussed thoroughly. First of all, it has been assumed that final data and performances—including durability—of Cheeses No. 1 and No. 2 are identical. In general, this assumption is questionable because 'old' cheeses (used in Cheese No. 2) tend to melt with good results, while 'young' cheeses (Cheese No. 1 formulation) tend to melt badly. On the other side, while protein molecules are demolished enough with a certain diminution when speaking of aqueous absorption [8]. Young cheeses tend to absorb more water than old cheeses. As a clear result, it should be assumed that Cheeses No. 1 and No. 2 have different chemical profiles and possibly very different microbiological profiles when speaking of possible spreading and accelerated perishability [6].

From the commercial and marketing angle, it has been often recognised that a particular into a well-defined category may be considered 'with high quality' if it has low durability. In other words, it may be sold at higher prices if its intrinsic durability is lower if compared with the mass of products into the above-mentioned category. This assumption is not always justifiable but allows to highlight a focal point when speaking of 'good prices' and 'claimed quality': the lower the durability for a food product, the higher the importance for this product from the commercial angle. Consequently, it may be assumed that the lower the durability, the higher the proposed price for a food, provided that all remaining conditions (including chemical and physical features, delivery prices) are unmodified.

On the other hand, it can be also assumed that the lower the RSL for a food product (the difference between the total shelf life and the consumed durability), the lower the possible price for the same item. This behaviour is always observable when speaking of commodities sold at lower prices when expiration dates or sell-by-date deadlines are close. Consequently, an inverse proportion between durability and price has to be considered, while the same price can be directly proportional to remaining durability.

On these bases, the simulation shown in Table 4.1 can be reviewed and amended assuming that:

(a) The amount of detectable water (moisture) in Cheeses No. 1 and 2 is not the same quantity.
(b) The amount of detectable moisture is predictably lower in Cheese No. 2 because of the use of 'old' cheeses.
(c) As a result, claimed durability is lower in Cheese No. 1 if compared with Cheese No. 2.
(d) Additionally, it has to be remembered that Cheese No. 1 is sold at 4.00 EUR and Cheese No. 2 is sold at 3.76 EUR.

Table 4.2 shows additional data for these formulations concerning calculated moisture, fat matter, protein content, pH, and the claimed durability (it has been calculated on the basis of two formulas for cheeses according to Parisi). Also, the determination of durability has been considered at $2 \pm 2\,°C$ and $10 \pm 2\,°C$ because of different storage options. Determination formulas for minimum durability according to Parisi need the calculation of moisture on free-fat basis (MFFB) amount, pH, and the count of yeasts and moulds [6]. The following data have been assumed also:

(1) 'Old' cheeses lose 20% of their moisture content (approximately 50%); this aqueous amount is completely lost.
(2) The count of yeasts and moulds is only 10 colony forming units (CFU)/g.
(3) pH values are 5.80 for Cheese No. 1 and 5.70 for Cheese No. 2.

As displayed in Table 4.2, Cheese No. 2 appears slightly less perishable than Cheese No. 1: 41 days at $2 \pm 2\,°C$ if compared with 39 days. Apparently, minimum durability at $10 \pm 2\,°C$ remains unchanged. In addition, the economic value (Table 4.1) is lower for Cheese No. 2 if compared with Cheese No. 1. As a result, the use of 'old' cheeses instead of 'young' ingredients has apparently determined

Table 4.2 Two different melted cheeses produced according to formulations shown in Table 4.1

Ingredients	Cheese No. 1	Cheese No. 2
	Amount	Amount
Moisture (%)	60.0	58.0
Fat matter (%)	22.0	22.0
Most reliable amount of proteins according to CYPEP:2006 (%)	15.1	16.8
Moisture on free-fat basis (MFFB) (%)	76.9	74.4
pH	5.80	5.70
Log_{10} yeast and mould count (CFU/g)	1.0	1.0
Minimum shelf-life value at 2 ± 2 °C (days)	39	41
Minimum shelf-life value at 10 ± 2 °C (days)	11	11
Economic value (EUR)	4.00	3.76

The use of 'old' cheeses in Cheese No. 2 has determined the reduction of water (moisture), the concomitant augment of fat and proteins, a slight diminution of pH, and the increase of minimum durability values at 2 ± 2 °C and 10 ± 2 °C. The final price for these cheeses has also been mentioned showing that the lower the claimed durability, the higher the possible price. Determination formulas for minimum durability according to Parisi need the calculation of moisture on free-fat basis (MFFB) amount, pH, and the count of yeasts and moulds [6, 10]

the augment of minimum durability performances with the concomitant decrease of prices.

Consequently, the simulation may be carried out with the aim of determining the best choices in terms of durability (low perishability) and low prices for processed cheeses. On the basis of displayed data for Cheese No. 1 and Cheese No. 2 (Tables 4.1 and 4.2), a food technologist might elaborate more than a single formulation where only the amount of 'old' cheeses is modified, as shown in Table 4.3.

In detail, Table 4.3 shows four hypothetical formulations, where

(a)   'Old' cheeses are comprised between 20% (Cheese No. 2) and 26% (No. 5).
(b)   Water decreases from 50 to 47%.
(c)   Vegetable fats decrease from 17% (No. 2) to 14% (No. 5).
(d)   Rennet casein decreases from 10% (No. 2) to 7% (No. 5).

As a result, the food technologist estimates four different production prices:

(1)   Cheese No. 2: 3.76 EUR;
(2)   Cheese No. 3: 3.63 EUR;
(3)   Cheese No. 4: 3.40 EUR;
(4)   Cheese No. 5: 3.36 EUR.

Table 4.4 shows additional data for these formulations concerning calculated moisture, fat matter, protein content, pH, and the claimed durability (it has been calculated on the basis of two formulas for cheeses according to Parisi). The determination of durability has been considered at 2 ± 2 °C and 10 ± 2 °C because of different storage options. Moreover, the following data have been assumed also:

Table 4.3   A simulation concerning four different processed cheeses

Ingredients	Cheese No. 2	Cheese No. 3	Cheese No. 4	Cheese No. 5
	Amount (%)		Amount (%)	
Water	50.0	49.0	48.0	47.0
Reused cheeses	20.0	22.0	24.0	26.0
Vegetable oil	17.0	16.0	14.5	14.0
Rennet casein	10.0	9.0	8.5	7.0
Food-grade salt	1.3	1.8	2.3	2.8
Melting agent(s) such as sodium citrate	1.2	1.5	1.8	2.0
Acidity corrector(s) such as citric acid	0.5	0.7	0.9	1.2
Economic value (EUR)	3.76	3.63	3.40	3.36

The quantity of 'old' cheeses increases from 20 (Cheese No. 2) to 26% (No. 5). Water decreases from 50 to 47%. Vegetable fats and rennet casein are modified. These formulations are hypothetical

Table 4.4   Four different melted cheeses produced according to formulations shown in Table 4.2

Ingredients	Cheese No. 2	Cheese No. 3	Cheese No. 4	Cheese No. 5
	Amount	Amount	Amount	Amount
Moisture (%)	58.0	57.8	57.6	57.4
Fat matter (%)	22.0	21.5	21.0	20.5
Most reliable amount of proteins according to CYPEP:2006 (%)	16.8	17.4	18.0	18.6
Moisture on free-fat basis (MFFB) (%)	74.4	73.6	72.9	72.2
pH	5.70	5.70	5.70	5.70
Log_{10} yeast and mould count (CFU/g)	1.0	1.0	1.0	1.0
Minimum shelf-life value at $2 \pm 2\,°C$ (days)	41	42	43	44
Minimum shelf-life value at $10 \pm 2\,°C$ (days)	11	12	12	13
Economic value (EUR)	3.76	3.63	3.40	3.36

The increase of 'old' cheeses has determined the reduction of water (moisture), the concomitant augment of fat and proteins, and the increase of minimum durability values at $2 \pm 2\,°C$ and $10 \pm 2\,°C$. The final price for these cheeses has also been mentioned showing that the lower the claimed durability, the higher the possible price. Determination formulas for minimum durability according to Parisi need the calculation of moisture on free-fat basis (MFFB) amount, pH, and the count of yeasts and moulds [6, 10]

(1) 'Old' cheeses lose always 20% of their moisture content (approximately 50%); this aqueous amount is completely lost.
(2) The count of yeasts and moulds is only 10 CFU/g.
(3) pH values remain constant (5.70).

In these conditions, the minimum durability of cheeses seems to increase at 2 ± 2 °C (41–44 days) and 10 ± 2 °C (11–13 days). Consequently, it may be inferred for these hypothetical formulations that:

(a) The increase of recycled foods has also augmented minimum durability (provided that other ingredients are modified with the aim of obtaining reliable products).
(b) In addition, the higher the amount of recycled foods, the lower the correspondent price.
(c) Consequently, prices appear inversely proportional to durability values, even if it has to be clarified that the independent variable is always the quantity of recycled cheeses.

Figure 4.1 shows the relationship between the increase of recycled cheeses and durability at 2 ± 2 °C, while Fig. 4.2 displays a similar relation between recycled cheeses and prices.

It should be also noted that prices may be calculated on the basis of durability data, even if the real independent variable is the amount of recycled cheeses. Figure 4.3 shows the relationship between prices and durability at 2 ± 2 °C: a polynomial interpolation of data ($R^2 = 0.9764$) seems to suggest that prices may be predictable on these bases.

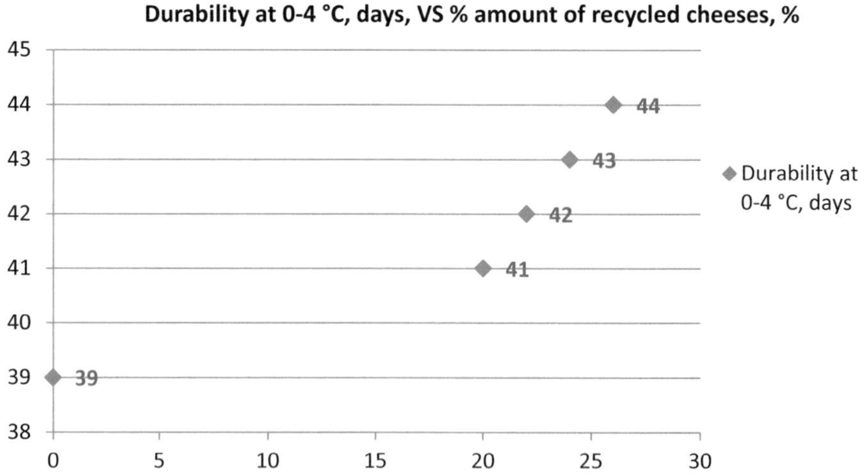

Durability at 0-4 °C, days, VS % amount of recycled cheeses, %

Fig. 4.1 Increase of minimum shelf-life values at 2 ± 2 °C for Cheeses No. 1–5, on the basis of the amount of recycled cheeses

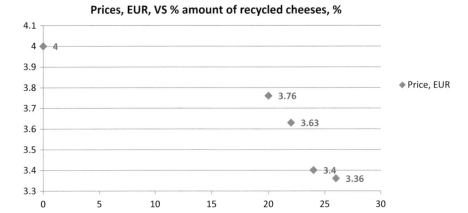

Fig. 4.2 Increase of price values for Cheeses No. 1–5, on the basis of the amount of recycled cheeses

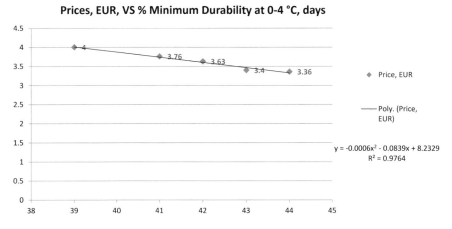

Fig. 4.3 Relationship between prices and durability at $2 \pm 2\,°C$ for Cheeses No. 1–5. A polynomial interpolation of data ($R^2 = 0.9764$) seems to suggest that prices may be predictable on these bases

4.4 Residual Durability, the Recycling Option… and the Best Choice

Above-mentioned and discussed simulations seem to demonstrate that the recycling of unused and still edible foods and beverages may be good enough when speaking of food waste countermeasures. However, it has to be highlighted that the main advantage is for food and beverage producers and the general market. In fact, food recycling simply allows for the reuse of unused cheeses (in our simulation) or foods and beverages, with the extension of shelf life for original food commodities. The reduction of prices has to be demonstrated; in addition, there is no demonstration that needful people can effectively take advantage of this system because food recycling occurs usually in a few and limited areas.

An indirect confirmation can be obtained by means of cheese simulations and the relationship shown in Fig. 4.2. On the one side, the increase of durability allows also for the augment of RSL. In other terms, the higher the total durability, the higher the RSL for a food product, provided that RSL is correctly defined. A possible formula could be RSL = minimum durability (at 2 ± 2 °C)/10 = 0.1 × minimum durability. Consequently, should minimum shelf life for simulated cheeses be 44 days, RSL would be $4.4 \cong 4$ days at 2 ± 2 °C. On a general level, RSL (intended as a fraction of total durability) tends to increase, with the obvious extension of food usability, for both food producers/manufacturers/packagers and food sharing organisations.

On the other hand, organisations such as social supermarkets tend to give unused foods at low—very low—prices if compared with the original value. As above-mentioned, price reductions should be 30–70%. Consequently, our simulated cheeses would be an important advantage in terms of food sharing only if the original price— 4.00 EUR—would be reduced at 2.80 EUR at least. A simple calculation based on Fig. 4.2 (price in function of recycled cheese) would demonstrate that this goal is possible only if 32.72% of the total formulation corresponds to recycled cheeses. This result is obtainable on condition that the first point (4.00 EUR vs. 0% of recycled cheeses) is eliminated. Otherwise, only a polynomial interpolation would be used giving a discouraging result (439.7% of recycled cheeses).

For this reason at least, food sharing should be considered as a distinctive advantage for food consumers on a general level. The recycle of unused and still eatable foods is certainly a good option [1] but could be also used as part of a more complex and multi-operational strategy with the fundamental action of food sharing organisations [19–29].

References

1. Lee P, Peter W, Hollins O (2010) Waste arisings in the supply of food and drink to households in the UK. Waste and Resources Action Programme, Banbury
2. Pekcan G, Köksal E, Küçükerdönmez Ö, Özel H (2006) Household food wastage in Turkey. Food and Agriculture Organization of the United Nations, Rome
3. Quested T, Johnson H (2009) Household food and drink waste in the UK. Waste and Resources Action Programme, Banbury
4. Mania I, Barone C, Caruso G, Delgado A, Micali M, Parisi S (2016) Traceability in the cheese making field. The regulatory ambit and practical solutions. Food Qual Manag 3:18–20 ISSN 2336-4602
5. Mania I, Delgado AM, Barone C, Parisi S (2018) Traceability in the dairy industry in Europe. Springer, Heidelberg
6. Parisi S (2002) Profili evolutivi dei contenuti batterici e chimico-fisici in prodotti lattiero-caseari. Ind Aliment 41(412):295–306
7. Lubeck J (2014) Viele Wege, Lebensmittel nicht z verschwenden. Die Welt. Available http://www.welt.de/regionales/hamburg/article132545750/Viele-WegeLebensmittel-nicht-zu-verschwenden.html. Accessed 29 May 2019

8. Parisi S, Laganà P, Delia AS (2006) Il calcolo indiretto del tenore proteico nei formaggi: il metodo CYPEP. Ind Aliment 462:997–1010
9. Parisi S, Laganà P, Delia AS (2007) Lo studio dei profili proteici durante la maturazione dei formaggi tramite il metodo CYPEP. Ind Aliment 468:404–417
10. Parisi S (2003) Evoluzione chimico-fisica e microbiologica nella conservazione di prodotti lattiero - caseari. Ind Aliment 423:249–259
11. Minsaas J, Heie AC (1979) Recycling of domestic food waste. Conserv Recycl 3(3–4):427–438. https://doi.org/10.1016/0361-3658(79)90040-74
12. Shaw P, Smith M, Williams I (2018) On the prevention of avoidable food waste from domestic households. Recycling 3(2):24. https://doi.org/10.3390/recycling3020024
13. Scholz K, Eriksson M, Strid I (2015) Carbon foot-print of supermarket food waste. Resour Conserv Recycl 94:56–65. https://doi.org/10.1016/j.resconrec.2014.11.016
14. Brancoli P, Rousta K, Bolton K (2017) Life cycle assessment of supermarket food waste. Resour Conserv Recycl 118:39–46. https://doi.org/10.1016/j.resconrec.2016.11.024
15. Refsgaard K, Magnussen K (2009) Household be-haviour and attitudes with respect to recycling food waste–experiences from focus groups. J Environ Manag 90(2):760–771. https://doi.org/10.1016/j.jenvman.2008.01.018
16. Fehr M, Calcado MDR, Romao DC (2002) The basis of a policy for minimizing and recycling food waste. Environ Sci Policy 5(3):247–253. https://doi.org/10.1016/S1462-9011(02)00036-9
17. Da Cruz NF, Ferreira S, Cabral M, Simoes P, Marques RC (2014) Packaging waste recycling in Europe: is the industry paying for it? Waste Manag 34(2):298–308. https://doi.org/10.1016/j.wasman.2013.10.035
18. Comber R, Thieme A (2013) Designing beyond habit: opening space for improved recycling and food waste behaviors through processes of persuasion, social influence and aversive affect. Pers Ubiquit Comput 17(6):1197–1210. https://doi.org/10.1007/s00779-012-0587-1
19. Fait M, Scorrano P, Mastroleo G, Cillo V, Scuotto V (2019) A novel view on knowledge sharing in the agri-food sector. J Knowl Manag. https://doi.org/10.1108/jkm-09-2018-0572
20. Shirado H, Iosifidis G, Tassiulas L, Christakis NA (2019) Resource sharing in technologically defined social networks. Nat Comm 10(1):1079. https://doi.org/10.1038/s41467-019-08935-2
21. Sidali KL, Kastenholz E, Bianchi R (2015) Food tourism, niche markets and products in rural tourism: combining the intimacy model and the experience economy as a rural development strategy. J Sustain Tour 23(8–9):1179–1197. https://doi.org/10.1080/09669582.2013.836210
22. Marovelli B (2019) Cooking and eating together in London: food sharing initiatives as collective spaces of encounter. Geoforum 99:190–201. https://doi.org/10.1016/j.geoforum.2018.09.006
23. Hamari J, Sjöklint M, Ukkonen A (2016) The sharing economy: why people participate in collaborative consumption. J Assoc Inf Sci Technol 67(9):2047–2059. https://doi.org/10.1002/asi.23552
24. Pascucci S, Dentoni D, Lombardi A, Cembalo L (2016) Sharing values or sharing costs? Understanding consumer participation in alternative food networks. NJAS Wageningen J Life Sci 78:47–60. https://doi.org/10.1016/j.njas.2016.03.006
25. Tornaghi C (2017) Urban agriculture in the food-disabling city: (Re)defining urban food justice. Reimagin Polit Empower Antipod 49(3):781–801. https://doi.org/10.1111/anti.12291
26. Falcone PM, Imbert E (2017) Bringing a sharing economy approach into the food sector: the potential of food sharing for reducing food waste. In: Morone P, Papendiek F, Tartiu V (eds) Food waste reduction and valorisation. Springer International Publishing, Cham. https://doi.org/10.1007/978-3-319-50088-1_10
27. Kaushal LA (2018) The Rise in the sharing economy: Indian perspective. In: Zakaria N, Kaushal LA (eds) Global entrepreneurship and new venture creation in the sharing economy. IGI Global, Hershey. https://doi.org/10.4018/978-1-5225-2835-7.ch007
28. Loh P, Agyeman J (2019) Urban food sharing and the emerging Boston food solidarity economy. Geoforum 99:213–222. https://doi.org/10.1016/j.geoforum.2018.08.017
29. Nishino N, Takenaka T, Takahashi H (2017) Manufacturer's strategy in a sharing economy. CIRP Ann 66(1):409–412. https://doi.org/10.1016/j.cirp.2017.04.004

Printed in the United States
By Bookmasters